A First Course in
Wavelets
with Fourier Analysis

A First Course in Wavelets with Fourier Analysis

Albert Boggess
Francis J. Narcowich
Texas A& M University, Texas

Prentice Hall

PRENTICE HALL, Upper Saddle River, NJ 07458

Library of Congress Cataloging-in-Publication Data

Boggess, Albert
A first course in wavelets with Fourier analysis / Albert Boggess, Francis J.
Narcowich.—1st ed.
    p.   cm.
Includes bibliographical references and index.
ISBN 0-13-022809-5
1. Wavelets (Mathematics) 2. Fourier analysis. I. Narcowich, Francis J. II. Title.

QA403.3 .B64 2001
512'.2433—dc21

00-066930

Acquisition Editor: George Lobell
Vice President/Director Production and Manufacturing: David W. Riccardi
Executive Managing Editor: Kathleen Schiaparelli
Senior Managing Editor: Linda Mihatov Behrens
Production Editor: Steven S. Pawlowski
Manufacturing Buyer: Alan Fischer
Manufacturing Manager: Trudy Pisciotti
Marketing Manager: Angela Battle
Marketing Assistant: Vince Jansen
Director of Marketing: John Tweeddale
Editorial Assistant: Melanie Van Benthuysen
Associate Editor, Mathematics/Statistics Media: Audra J. Walsh
Art Director: Jayne Conte
Cover Designer: Bruce Kenselaar
Cover Image: Marita Froimson

 ©2001 by Prentice-Hall, Inc.
Upper Saddle River, New Jersey 07458

Printed in the United States of America

10   9   8   7   6   5   4   3   2

ISBN 0-13-022809-5

Prentice-Hall International (UK) Limited, London
Prentice-Hall of Australia Pty. Limited, Sydney
Prentice-Hall Canada Inc., Toronto
Prentice-Hall Hispanoamericana, S. A., Mexico
Prentice-Hall of India Private Limited, New Delhi
Prentice-Hall of Japan, Inc., Tokyo
Pearson Education Asia Pte. Ltd.
Editora Prentice-Hall do Brasil, Ltda., Rio de Janeiro

The authors dedicate this book to
Eugene Wigner and Mary Boas

Contents

Preface

Fourier series and the Fourier transform have been around since the nineteenth century and many research articles and books (at both the graduate and undergraduate levels) have been written about these topics. By contrast, the development of wavelets has been much more recent. While its origins go back many decades, the subject of wavelets has become a popular tool in signal analysis and other areas of applications only within the last two decades or so partly as a result of Ingrid Daubechies's celebrated work on the construction of compactly supported, orthonormal wavelets. Consequently, most of the articles and reference materials on wavelets require a sophisticated mathematical background (a good first-year real analysis course at the graduate level). Our goal with this book is to present many of the essential ideas behind Fourier analysis and wavelets, along with some of their applications to signal analysis, to an audience of advanced undergraduate science, engineering, and mathematics majors. The only prerequisites are a good calculus background and some exposure to linear algebra (a course that covers matrices, vector spaces, linear independence, linear maps, and inner product spaces should suffice). The applications to signal processing are kept elementary, without much use of the technical jargon of the subject, in order for this material to be accessible to a wide audience.

Fourier Analysis

The basic goal of Fourier series is to take a signal, which will be considered as a function of the time variable t, and decompose it into its various frequency components. The basic building blocks are the sine and cosine functions:

$$\sin(nt) \qquad \cos(nt),$$

which vibrate at a frequency of n times per 2π interval. As an example, consider the following function:

$$f(t) = \sin(t) + 2\cos(3t) + 0.3\sin(50t).$$

This function has three components that vibrate at frequency 1 (the $\sin t$ part), at frequency 3 [the $2\cos(3t)$ part], and at frequency 50 [the $0.3\sin(50t)$ part]. The graph of f is given in Figure 1.

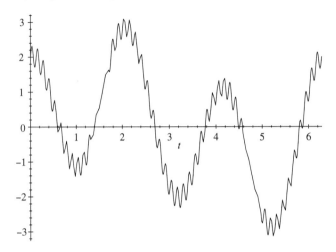

Figure 1 Plot of $f(t) = \sin(t) + 2\cos(3t) + 0.3\sin(50t)$

A common problem in signal analysis is to filter out unwanted noise. The background hiss on a cassette tape is an example of high-frequency (audio) noise that various devices (Dolby filters) try to filter out. In the preceding example, the component, $0.3\sin(50t)$, contributes the high-frequency wiggles to the graph of f in Figure 1. By setting the coefficient 0.3 equal to zero, the resulting function is

$$\tilde{f}(t) = \sin(t) + 2\cos(3t)$$

whose graph (given in Figure 2) is the same as the one for f but without the high-frequency wiggles.

The preceding example shows that one approach to the problem of filtering out unwanted noise is to express a given signal, $f(t)$, in terms of sines and cosines:

$$f(t) = \sum_n a_n \cos(nt) + b_n \sin(nt)$$

and then to eliminate (i.e., set equal to zero) the coefficients (the a_n and b_n) that correspond to the unwanted frequencies. In the case of the signal f just presented, this process is easy since the signal is already presented as a sum of sines and cosines. Most signals, however, are not presented in this manner. The subject of Fourier series, in part, is the study of how to efficiently decompose a function into a sum of cosine and sine components so that various types of filtering can be accomplished easily.

Another related problem in signal analysis is that of data compression. Imagine that the graph of the signal $f(t)$ in Figure 1 represents a telephone conversation. The horizontal axis is time, perhaps measured in milliseconds, and the

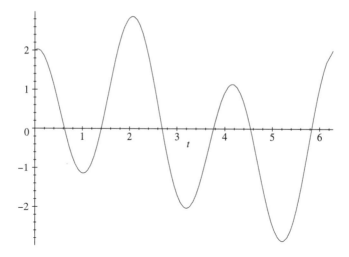

Figure 2 Plot of $f(t) = \sin(t) + 2\cos(3t)$

vertical axis represents the electric voltage of a sound signal generated by someone's voice. Suppose this signal is to be digitized and sent via satellite overseas from America to Europe. One naive approach is to sample the signal every millisecond or so and send these data bits across the Atlantic. However, this would result in thousands of data bits per second for just one phone conversation. Since there will be many such conversations between the two continents, the phone company would like to compress this signal into as few digital bits as possible without significantly distorting the signal. A more efficient approach is to express the signal into its Fourier series: $f(t) = \sum_n a_n \cos(nt) + b_n \sin(nt)$ and then discard those coefficients, a_n and b_n, that are smaller than some tolerance for error. Only those coefficients that are above this tolerance need to be sent across the Atlantic, where the signal can then be reconstructed. For most signals, the number of significant coefficients in its Fourier series is relatively small.

Wavelets

One disadvantage of Fourier series is that its building blocks, sines and cosines, are periodic waves that continue forever. While this approach may be appropriate for filtering or compressing signals that have time-independent wavelike features (as in Figure 1), other signals may have more localized features for which sines and cosines do not model very well. As an example, consider the graph given in Figure 3. This may represent a sound signal with two isolated noisy pops that need to be filtered out. Since these pops are isolated, sines and cosines do not model this signal very well. A different set of building blocks, called *wavelets*, is designed to model these types of signals. In a rough sense, a wavelet looks like a wave that travels for one or more periods and is nonzero only

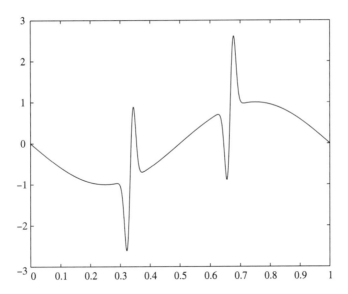

Figure 3 Graph of a signal with isolated noise

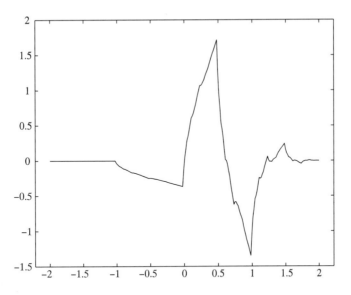

Figure 4 Graph of Daubechies wavelet

over a finite interval instead of propagating forever the way sines and cosines do [see Figure 4 for the graph of the Daubechies ($N = 2$) wavelet]. A wavelet can be translated forward or backward in time. It also can be stretched or compressed by scaling to obtain low- and high-frequency wavelets (see Figure 5). Once a wavelet function is constructed, it can be used to filter or compress signals in

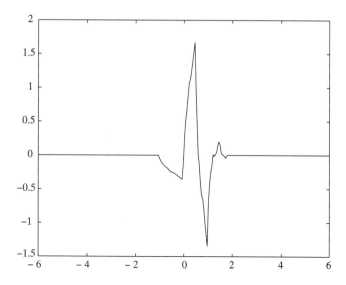

Figure 5 High-frequency Daubechies wavelet

much the same manner as Fourier series. A given signal is first expressed as a sum of translations and scalings of the wavelet. Then the coefficients corresponding to the unwanted terms are removed or modified.

In order to implement efficient algorithms for decomposing a signal into an expansion (either Fourier or wavelet based), the building blocks (sines, cosines or wavelets) should satisfy various properties. One convenient property is *orthogonality*, which for the sine function states

$$\frac{1}{\pi} \int_0^{2\pi} \sin(nt)\sin(mt)\, dt = \begin{cases} 0 & \text{if } n \neq m \\ 1 & \text{if } n = m. \end{cases}$$

The analogous properties hold for the cosine function as well. In addition, $\int_0^{2\pi} \sin(nt)\cos(mt)\, dt = 0$ for all n and m. We shall see that these orthogonality properties result in simple formulas for the Fourier coefficients (the a_n and b_n) and efficient algorithms (fast Fourier transform) for their computation.

One of the difficult tasks in the construction of a wavelet is to make sure that its translates and rescalings satisfy analogous orthogonality relationships, so that efficient algorithms for the computation of the wavelet coefficients of a given signal can be found. This is why we cannot construct a wavelet simply by truncating a sine or cosine wave by declaring it to be zero outside of one or more of its periods. Such a function, while satisfying the desired support feature of a wavelet, would not satisfy any reasonable orthogonality relationship with its translates and rescales and thus would not be as useful for signal analysis.

Outline

This text has eight chapters and two appendices. Chapter 0, on inner product spaces, contains the necessary prerequisites for Chapters 1 through 7. The primary inner product space of interest is the space of square integrable functions, which is presented in simplified form without the use of the Lebesgue integral. Depending on the audience, this chapter can be covered at the beginning of a course or can be folded into the course as the need arises. Chapter 1 contains the basics of Fourier series. Several convergence theorems are presented with simplifying hypothesis so that their proofs are manageable. The Fourier transform is presented in Chapter 2. Besides being of interest in its own right, much of this material is used in later chapters on wavelets. An informal proof of the Fourier inversion formula is presented in order to keep the exposition at an elementary level. A formal proof is given in the Appendix A. The discrete Fourier transform and fast Fourier transform are discussed in Chapter 3. This chapter also contains applications to signal analysis and to the identification of the natural vibrating frequency (or sway) of a building.

Wavelets are discussed in Chapters 4 through 7. Our presentation on wavelets starts with the case of the Haar wavelets in Chapter 4. The basic ideas behind a multiresolution analysis and the desired features of wavelets, such as orthogonality, are easy to describe with the explicitly defined Haar wavelets. However, the Haar wavelets are discontinuous and so they are of limited use in signal analysis. The concept of a multiresolution analysis in a general context is presented in Chapter 5. This gives a general framework that generalizes the structure of the wavelet spaces generated by the Haar wavelet. Chapter 6 contains the construction of the Daubechies wavelet, which is continuous and orthogonal. Prescriptions for smoother wavelets are also given. Chapter 7 contains more advanced topics, such as wavelets in higher dimensions and the wavelet transform.

The proofs of most theorems are given in the text. Some of the more technical theorems are discussed in a heuristic manner with complete proofs given in Appendix A. Some of these proofs require more advanced mathematics, such as some exposure to the Lebesgue integral.

MATLAB code that was used to generate figures or to illustrate concepts is found in Appendix B.

This text is not a treatise. The focus of the latter half of the book is on the construction of orthonormal wavelets. Little mention is made of bi-orthogonal wavelets using splines and other tools. There are ample references for these other types of wavelets (see, for example, [5]) and we want to keep the amount of material in this text manageable for a one-semester undergraduate course.

The basics of Fourier analysis and wavelets can be covered in a one semester undergraduate course using the following outline:

- Chapter 0, Sections 0.1 through 0.5 (Sections 0.6 and 0.7 on adjoints, least squares, and linear predictive coding are more topical in nature). This material can either be covered first or covered as needed throughout the rest of the course.

- Chapter 1 (Fourier Series), all sections.

- Chapter 2 (Fourier Transform), all sections except the ones on the adjoint of the Fourier transform, and the proof of the uncertainty principle, which are more topical in nature.

- Chapter 3 (Discrete Fourier Analysis), all sections except the Z-transform, which is more topical in nature.

- Chapter 4 (Haar Wavelet Analysis), all sections.

- Chapter 5 (Multiresolution Analysis), all sections.

- Chapter 6 (Daubechies Wavelets), all sections.

Acknowledgments

This book arose from lecture notes used by both authors for the Fourier Analysis and Wavelets course taught at Texas A&M. The authors would like to thank the many students in these classes that gave critical comments on the manuscript. We would especially like to thank Svenja Lowitzsch and Beng Ong who read the manuscript carefully and corrected many of our mistakes. The authors would also like to thank the editorial staff; in particular, George Lobell, for the kind encouragement during the development of this book. the following reviewers did an excellent job of making suggestions and pointing out errors, William Beckner, University of Texas; Joe Lakey, New Mexico State University; Edward A. Newburg, Rochester Institute of Technology; Oscar Rothaus, Cornell University; and David Weinberg, Texas Tech university. Of course, any mistakes remaining in the book are solely the fault of the authors. We also thank Steven S. Pawlowski and the rest of the staff at Prentice Hall for their professional production of the book. On a personal note, Fran Narcowich would also like to thank his wife Linda for support and encouragement.

Albert Boggess
Al.Boggess@math.tamu.edu

Francis J. Narcowich
fnarc@math.tamu.edu

A First Course in
Wavelets
with Fourier Analysis

Chapter 0

Inner Product Spaces

0.1 Motivation

For two vectors $X = (x_1, x_2, x_3)$, $Y = (y_1, y_2, y_3)$ in R^3, the standard (Euclidean) inner product of X and Y is defined as

$$\langle X, Y \rangle = x_1 y_1 + x_2 y_2 + x_3 y_3.$$

This definition is partly motivated by the desire to measure the length of a vector, which is given by the Pythagorean theorem:

$$\text{Length of } X = \sqrt{x_1^2 + x_2^2 + x_3^2} = \sqrt{\langle X, X \rangle}.$$

The goal of this chapter is to define the concept of an inner product in a more general setting that includes a wide variety of vector spaces. We are especially interested in the inner product defined on vector spaces whose elements are signals (i.e., functions of time).

0.2 Definition of Inner Product

The definition of an inner product in R^3 naturally generalizes to R^n for any dimension n. For two vectors $X = (x_1, x_2, \cdots, x_n)$, $Y = (y_1, y_2, \cdots, y_n)$ in R^n, the Euclidean inner product is

$$\langle X, Y \rangle = \sum_{j=1}^{n} x_j y_j.$$

1

When we study Fourier series and the Fourier transform, we will use heavily the complex exponential. Thus, we must consider complex vector spaces as well as real ones. The preceding definition of the inner product for R^n can be modified for vectors in C^n by conjugating the second factor. Recall that the conjugate of a complex number $z = x + iy$ is defined as $\bar{z} = x - iy$. Note that $z\bar{z} = x^2 + y^2$, which by definition is $|z|^2$ [the square of the length of $z = x + iy$ regarded as vector in the plane from $(0,0)$ to (x, y)].

If $Z = (z_1, z_2, \cdots, z_n)$, $W = (w_1, w_2, \cdots, w_n)$ are vectors in C^n, then

$$\langle Z, W \rangle = \sum_{j=1}^{n} z_j \overline{w_j}.$$

The purpose of the conjugate is to ensure that the length of a vector in C^n is real and nonnegative:

$$\text{Length of } Z = \sqrt{\langle Z, Z \rangle} = \sqrt{\sum_{j=1}^{n} z_j \overline{z_j}} = \sqrt{\sum_{j=1}^{n} |z_j|^2}.$$

The inner products just defined share certain properties. For example, the inner product is bilinear, which implies

$$\langle X + Y, Z \rangle = \langle X, Z \rangle + \langle Y, Z \rangle \quad \text{and} \quad \langle X, Y + Z \rangle = \langle X, Y \rangle + \langle X, Z \rangle.$$

The rest of the properties satisfied by the aforementioned inner products are set down as axioms in the following definition. We shall leave the verification of these axioms for the inner products for R^n and C^n as exercises.

DEFINITION 0.1 *An inner product on a complex vector space V is a function $\langle \cdot, \cdot \rangle : V \times V \to C$ that satisfies the following properties.*

- *Positivity: $\langle v, v \rangle > 0$ for each nonzero $v \in V$.*

- *Conjugate symmetry: $\overline{\langle v, w \rangle} = \langle w, v \rangle$ for all vectors v and w in V.*

- *Homogeneity: $\langle cv, w \rangle = c\langle v, w \rangle$ for all vectors v and w in V and scalars $c \in C$.*

- *Additivity: $\langle u + v, w \rangle = \langle u, w \rangle + \langle v, w \rangle$ for all u, v, $w \in V$.*

A vector space with an inner product is called an inner product space.

To emphasize the underlying space V, we shall sometimes denote the inner product on V by

$$\langle \ , \ \rangle_V.$$

The preceding definition also serves to define a real inner product on a real vector space except that the scalar c in the homogeneity property is real and there is no conjugate in the statement of conjugate symmetry.

Note that the second and fourth properties imply bilinearity in the second factor: $\langle u, v + w \rangle = \langle u, v \rangle + \langle u, w \rangle$. The second and third properties imply that scalars factor out of the second factor with a conjugate:

$$\langle v, cw \rangle = \overline{\langle cw, v \rangle} = \overline{c}\,\overline{\langle w, v \rangle} = \overline{c}\langle v, w \rangle.$$

The positivity condition means that we can assign the nonzero number, $\|v\| = \sqrt{\langle v, v \rangle}$, as the *length* or *norm* of the vector v. The notion of length gives meaning to the distance between two vectors in V, by declaring

$$\text{Distance between } \{v, w\} = \|v - w\|.$$

Note that the positivity property of the inner product implies that the only way $\|v - w\| = 0$ is when $v = w$. This notion of distance also gives meaning to the idea of a convergent sequence $\{v_k; \ k = 1, 2, \dots\}$; namely, we say that

$$v_k \to v \quad \text{if} \quad \|v_k - v\| \to 0.$$

In words, $v_k \to v$ if the distance between v_k and v gets small as k gets large.

Here are some further examples of inner products.

EXAMPLE 0.2

Let V be the space of polynomials $p = a_n x^n + \cdots + a_1 x + a_0$, with $a_j \in C$. An inner product on V is given as follows: if $p = a_0 + a_1 x + \cdots + a_n x^n$ and $q = b_0 + b_1 x + \cdots + b_n x^n$, then

$$\langle p, q \rangle = \sum_{j=0}^{n} a_j \overline{b_j}.$$

Note that this inner product space looks very much like C^{n+1}, where we identify a point $(a_0, \dots, a_n) \in C^{n+1}$ with $a_0 + a_1 x + \cdots + a_n x^n$. ∎

EXAMPLE 0.3

Different inner products can be imposed on the same vector space. This example defines an inner product on C^2 that is different from the standard Euclidean inner product. Suppose $v = (v_1, v_2)$ and $w = (w_1, w_2)$ are vectors in C^2. Define

$$\langle v, w \rangle = (\overline{w_1}, \overline{w_2}) \begin{pmatrix} 2 & -i \\ i & 3 \end{pmatrix} \begin{pmatrix} v_1 \\ v_2 \end{pmatrix} \quad \text{(ordinary matrix multiplication).} \quad ∎$$

There is nothing special about the particular choice of matrix. We can replace the matrix in the preceding equation with any matrix A as long as it is Hermitian symmetric (i.e. $\overline{A}^T = A$) and positive definite (i.e. all eigenvalues are positive). These conditions imply that A is invertible. Verification of these properties will be left as exercises.

0.3 The Spaces L^2 and l^2

0.3.1 Definitions

The examples in the last section are all finite dimensional (i.e., they contain only a finite number of linearly independent vectors). In this section, we discuss a class of infinite dimensional vector spaces that is particularly useful for analyzing signals. A signal (for example, a sound signal) can be viewed as a function, $f(t)$, that indicates the intensity of the signal at time t. Here t varies in an interval $a \leq t \leq b$ that represents the time duration of the signal. Here, a could be $-\infty$ or b could be $+\infty$.

We will need to impose a growth restriction on the functions defined on the interval $a \leq t \leq b$. This leads to the following definition.

DEFINITION 0.4 *For an interval $a \leq t \leq b$, the space $L^2([a,b])$ is the set of all square integrable functions defined on $a \leq t \leq b$. In other words,*

$$L^2([a,b]) = \left\{ f : [a,b] \to C; \ \int_a^b |f(t)|^2 \, dt < \infty \right\}.$$

Functions that are discontinuous are allowed as members of this space. All the examples considered in this book are either continuous or discontinuous at a finite set of points. In this context, the preceding integral can be interpreted in the elementary Riemann sense (the one introduced in freshmen calculus courses). The definition of L^2 allows functions whose set of discontinuities is quite large, in which case the Lebesgue integral must be used. The condition $\int_a^b |f(t)|^2 \, dt < \infty$ physically means that the total energy of the signal is finite (which is a reasonable class of signals to consider).

The space $L^2[a,b]$ is infinite dimensional. For example, if $a = 0$ and $b = 1$, then the set of functions $\{1, \ t, \ t^2, \ t^3 \dots \}$ is linearly independent and belongs to $L^2[0,1]$. The function $f(t) = 1/t$ is an example of a function that does not belong to $L^2[0,1]$ since $\int_0^1 (1/t)^2 \, dt = \infty$.

L^2 **Inner Product.** We now turn our attention to constructing an appropriate inner product on $L^2[a,b]$. To motivate the L^2 inner product, we discretize the interval $[a,b]$. To simplify matters, let $a = 0$ and $b = 1$. Let N be a large positive integer and let $t_j = j/N$ for $1 \leq j \leq N$. If f is continuous, then the values of f on the interval $[t_j, t_{j+1})$ can be approximated by $f(t_j)$. Therefore, f can be approximated by the following vector:

$$f_N = (f(t_1), f(t_2), \dots, f(t_N)) \in R^N$$

as illustrated in Figure 1. As N gets larger, f_N becomes a better approximation to f.

If f and g are two signals in $L^2[0,1]$, then both signals can be discretized as f_N and g_N. One possible definition of $\langle f, g \rangle_{L^2}$ is to examine the ordinary R^N

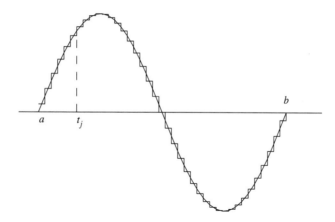

Figure 1 Approximating a continuous function by discretization

inner product of f_N and g_N as N gets large:

$$\langle f_N, g_N \rangle_{R^N} = \sum_{j=1}^{N} f(t_j)\overline{g(t_j)} = \sum_{j=1}^{N} f(j/N)\overline{g(j/N)}.$$

The trouble with this approach is that as N gets large, the sum on the right typically gets large. A better choice is to consider the averaged inner product:

$$\frac{1}{N}\langle f_N, g_N \rangle_{R^N} = \sum_{j=1}^{N} f(j/N)\overline{g(j/N)}\,\frac{1}{N}.$$

Since f_N and g_N approach f and g as N gets large, a reasonable definition of $\langle f, g \rangle_{L^2}$ is to take the limit of this averaged inner product as $N \to \infty$.

The preceding equation can be written as

$$\frac{1}{N}\langle f_N, g_N \rangle_{R^N} = \sum_{j=1}^{N} f(t_j)\overline{g(t_j)}\,\Delta t \quad \text{with } \Delta t = 1/N.$$

The sum on the right is a Riemann sum approximation to $\int_0^1 f(t)\overline{g(t)}\,dt$ over the partition $[0, t_1, t_2, \ldots, t_N = 1]$ of $[0, 1]$. This approximation gets better as N gets larger. Thus, a reasonable definition of an inner product on $L^2[0, 1]$ is $\langle f, g \rangle = \int_0^1 f(t)\overline{g(t)}\,dt$. This motivation provides the basis for the following definition.

DEFINITION 0.5 *The L^2 inner product on $L^2([a, b])$ is defined as*

$$\langle f, g \rangle_{L^2} = \int_a^b f(t)\overline{g(t)}\,dt \quad \text{for } f,\, g \in L^2([a, b]).$$

The conjugate symmetry, homogeneity, and bilinearity properties are all easily established for this inner product and we leave them as exercises.

For the positivity condition, if $0 = \langle f, f \rangle = \int_a^b |f(t)|^2\, dt$ and if f is continuous, then $f(t) = 0$ for all t (see Exercise 4). If $f(t)$ is allowed to be discontinuous at a finite number of points, then we can only conclude that $f(t) = 0$ at all but a finite number of t-values. For example, the function

$$f(t) = \begin{cases} 1 & \text{if } t = 0 \\ 0 & \text{otherwise} \end{cases}$$

is not the zero function yet $\int_{-1}^1 |f(t)|^2\, dt = 0$. However, we stipulate that two elements f and g in $L^2([a, b])$ are equal if $f(t) = g(t)$ for all values t except for a finite number of t-values (or, more generally, a set of measure zero if the Lebesgue integral is used). This is a reasonable definition for the purposes of integration, since $\int_a^b f(t)\, dt = \int_a^b g(t)\, dt$ for such functions. With this convention, the positivity condition holds.

This notion of equivalence is reasonable from the point of view of signal analysis. The behavior of a signal at one instant in time (say $t = 0$) is rarely important. The behavior of a signal over a time interval of positive length *is* important. Although measure theory and the Lebesgue integral are not used in this text, we digress to discuss this topic just long enough to put the notion of equivalence discussed in the previous paragraph in a broader context. The concept of measure of a set generalizes the concept of length of an interval. The measure of an interval $a < t < b$ is $b - a$. The measure of a disjoint union of intervals is the sum of their lengths. So the measure of a finite (or countably infinite) set of points is zero. The measure of a more complicated set can be determined by decomposing it into a limit of sets that are disjoint unions of intervals. Since intervals of length zero have no effect on integration, it is reasonable to expect that if a function f is zero on $a \le t \le b$ except on a set of measure zero, then $\int_a^b f(t)\, dt = 0$. The converse is also true: If

$$0 = \|f\|_{L^2[a,b]}^2 = \int_a^b f(t)^2\, dt,$$

then $f(t) = 0$ on $a \le t \le b$ except possibly on a set of measure zero. For this reason, it is reasonable to declare that two functions, f and g, are equivalent on $[a, b]$ if $f(t) = g(t)$ for all t in $[a, b]$ except possibly for a set of measure zero. This general notion of equivalence includes the definition stated in the previous paragraph (that two functions are equivalent if they agree except at a finite number of points).

The Space l^2. For many applications, the signal is already discrete. For example, the signal from a compact disc player can be represented by a discrete set of numbers that represent the intensity of its sound signal at regular (small) time intervals. In such cases, we represent the signal as a sequence

$X = \ldots, x_{-1}, x_0, x_1, \ldots$, where each x_j is the numerical value of the signal at the jth time interval $[t_j, t_{j+1}]$. Theoretically, the sequence could continue indefinitely (either as $j \to \infty$ or as $j \to -\infty$ or both). In reality, the signal usually stops after some point, which mathematically can be represented by $x_j = 0$ for $|j| > N$ for some integer N.

The following definition describes a discrete analogue of L^2.

DEFINITION 0.6 *The space l^2 is the set of all sequences $X = \ldots, x_{-1}, x_0, x_1,$ $\ldots$, $x_i \in C$, with $\sum_{-\infty}^{\infty} |x_n|^2 < \infty$. The inner product on this space is defined as*

$$\langle X, Y \rangle_{l^2} = \sum_{n=-\infty}^{\infty} x_n \overline{y_n}$$

for $X = \ldots, x_{-1}, x_0, x_1, \ldots$, and $Y = \ldots, y_{-1}, y_0, y_1, \ldots$.

Verifying that $\langle \cdot, \cdot \rangle$ is an inner product for l^2 is relatively easy and will be left to the exercises.

Relative Error. For two signals, f and g, the L^2-norm of their difference, $\|f - g\|_{L^2}$, provides one way of measuring how f differs from g. However, often the *relative error* is more meaningful:

$$\text{Relative error} = \frac{\|f - g\|_{L^2}}{\|f\|_{L^2}}$$

(the denominator could also be $\|g\|_{L^2}$). The relative error measures the L^2-norm of the difference between f and g in relation to the size of $\|f\|_{L^2}$. For discrete signals, the l^2-norm is used.

0.3.2 Convergence in L^2 versus Uniform Convergence

As defined in Section 0.2, a sequence of vectors $\{v_n;\ n = 1, 2, \ldots\}$ in an inner product space V is said to *converge* to the vector $v \in V$ provided that v_n is close to v when n is large. Closeness here means that $\|v_n - v\|$ is small. To be more mathematically precise, v_n converges to v if $\|v_n - v\| \to 0$ as $n \to \infty$.

In this text, we will often deal with the inner product space $L^2[a, b]$, and therefore we shall discuss convergence in this space in more detail.

DEFINITION 0.7 *A sequence f_n converges to f in $L^2[a, b]$ if $\|f_n - f\|_{L^2} \to 0$ as $n \to \infty$. More precisely, given any tolerance $\epsilon > 0$, there exists a postive integer N such that if $n \geq N$, then $\|f - f_n\|_{L^2} < \epsilon$.*

Convergence in L^2 is sometimes called *convergence in the mean*. There are two other types of convergence often used with functions.

DEFINITION 0.8

1. *A sequence f_n converges to f pointwise on the interval $a \leq t \leq b$ if for each $t \in [a, b]$ and each small tolerance $\epsilon > 0$, there is a positive integer N such that if $n \geq N$, then $|f_n(t) - f(t)| < \epsilon$.*

2.   *A sequence f_n converges to f uniformly on the interval $a \leq t \leq b$ if for each small tolerance $\epsilon > 0$, there is a positive integer N such that if $n \geq N$, then $|f_n(t) - f(t)| < \epsilon$ for all $a \leq t \leq b$.*

For uniform convergence, the N only depends on the size of the tolerance ϵ and not on the point t, whereas for pointwise convergence, the N is also allowed to depend on the point t.

How do these three types of convergence compare? If f_n uniformly converges to f on $[a, b]$, then the values of f_n are close to the values of f over the entire interval $[a, b]$. For example, Figure 2 illustrates the graphs of two functions that are uniformly close to each other. By contrast, if f_n converges to f pointwise, then for each fixed t, $f_n(t)$ is close to $f(t)$ for large n. However, the rate at which $f_n(t)$ approaches $f(t)$ may depend on the point t. Thus, a sequence that converges uniformly also must converge pointwise, but not conversely.

Figure 2 Graphs of two functions that are uniformly close

If f_n converges to f in $L^2[a, b]$ then on average, f_n is close to f, but for some values $f_n(t)$ may be far away from $f(t)$. For example, Figure 3 illustrates two functions that are close in L^2 even though some of their function values are not close.

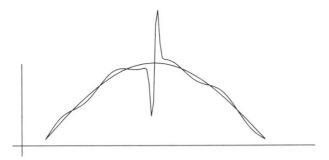

Figure 3 Two graphs that are close in L^2 but not uniformly close

EXAMPLE 0.9

The sequence of functions $f_n(t) = t^n$, $n = 1, 2, 3 \ldots$ converges pointwise to $f(t) = 0$ on the interval $0 \leq t < 1$ because for any number $0 \leq t < 1$, $t^n \to 0$

as $n \to \infty$. However, the convergence is not uniform. The rate at which t^n approaches zero becomes slower as t approaches 1. For example, if $t = 1/2$ and $\epsilon = 0.001$, then $|t^n| < \epsilon$ provided that $n \geq 10$. However, if $t = 0.9$, then $|t^n|$ is not less than ϵ until $n \geq 66$.

For any fixed number $r < 1$, then f_n converges uniformly to $f = 0$ on the interval $[0, r]$. Indeed, if $0 \leq t \leq r$, then $|t^n| \leq r^n$. Therefore, as long as r^n is less than ϵ, $|f_n(t)|$ will be less than ϵ for all $0 \leq t \leq r$. In other words, the rate at which f_n approaches zero for all points on the interval $[0, r]$ is no worse than the rate at which r^n approaches zero.

We also note that $f_n \to 0$ in $L^2[0, 1]$ because

$$\|f_n\|_{L^2}^2 = \int_0^1 (t^n)^2 \, dt$$

$$= \frac{t^{2n+1}}{2n + 1} \Big|_0^1$$

$$= \frac{1}{2n + 1} \to 0 \quad \text{as } n \to \infty. \qquad \blacksquare$$

As the following theorem shows, uniform convergence on a finite interval $[a, b]$ is a stronger type of convergence than L^2 convergence.

THEOREM 0.10 *If a sequence f_n converges uniformly to f as $n \to \infty$ on a finite interval $a \leq t \leq b$, then this sequence also converges to f in $L^2[a, b]$. The converse of this statement is not true.*

Proof Using the definition of uniform convergence, we can choose, for a given tolerance $\epsilon > 0$, an integer N such that

$$|f_n(t) - f(t)| < \epsilon \quad \text{for} \quad n \geq N \quad \text{and} \quad a \leq t \leq b.$$

This inequality implies

$$\|f_n - f\|_{L^2}^2 = \int_a^b |f_n(t) - f(t)|^2 \, dt$$

$$\leq \int_a^b \epsilon^2 \, dt \quad \text{for} \quad n \geq N$$

$$= \epsilon^2 (b - a).$$

Therefore, if $n \geq N$, we have $\|f_n - f\|_{L^2} \leq \epsilon \sqrt{b - a}$. Since ϵ can be chosen as small as desired, this inequality implies that f_n converges to f in L^2.

To show that the converse is false, consider the following sequence of functions on $0 \leq t \leq 1$:

$$f_n(t) = \begin{cases} 1 & 0 < t \leq 1/n \\ 0 & \text{otherwise.} \end{cases}$$

We leave it to the reader (see Exercise 6) to show that this sequence converges to the zero function in $L^2[0,1]$ but does not converge to zero uniformly on $0 \leq t \leq 1$.   ♦

In general, a sequence that converges pointwise does not necessarily converge in L^2. However, if the sequence is uniformly bounded by a fixed function in L^2, then pointwise convergence is enough to guarantee convergence in L^2 (this is the Lebesgue dominated convergence theorem; see [17]). Further examples illustrating the relationships between these three types of convergence are developed in the Exercises.

0.4   Schwarz and Triangle Inequalities

The two most important properties of inner products are the Schwarz and triangle inequalities. The Schwarz inequality states that $|\langle X, Y \rangle| \leq \|X\| \|Y\|$. In R^3, this inequality follows from the law of cosines:

$$|\langle X, Y \rangle| = \|X\| \|Y\| |\cos(\theta)| \leq \|X\| \|Y\|$$

where θ is the angle between X and Y. The triangle inequality states that $\|X + Y\| \leq \|X\| + \|Y\|$. In R^3, this inequality follows from Figure 4 which expresses the fact that the shortest distance between two points is a straight line.

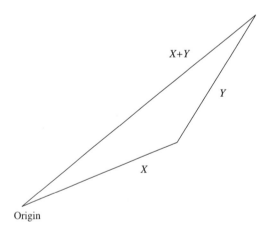

Figure 4 Triangle inequality

The following theorem states that the Schwarz and triangle inequalities hold for general inner product spaces.

THEOREM 0.11 *Suppose V, $\langle \cdot, \cdot \rangle$ is an inner product space (either real or complex). Then for all $X, Y \in V$,*

- Schwarz inequality: $|\langle X, Y \rangle| \le \|X\| \, \|Y\|$. *Equality holds if and only if X and Y are linearly dependent. Moreover, $\langle X, Y \rangle = \|X\| \, \|Y\|$ if and only if X or Y is a nonnegative multiple of the other.*

- Triangle inequality: $\|X + Y\| \le \|X\| + \|Y\|$. *Equality holds if and only if X or Y is a nonnegative multiple of the other.*

Proof for Real Inner Product Spaces

Assume that one of the vectors, say Y, is nonzero, for otherwise there is nothing to show. Let t be a real variable and consider the following inequality:

$$0 \le \|X - tY\|^2 = \langle X - tY, X - tY \rangle \tag{0.1}$$

$$= \|X\|^2 - 2t \langle X, Y \rangle + t^2 \|Y\|^2. \tag{0.2}$$

The right side is a nonnegative quadratic polynomial in t, and so it cannot have two distinct real roots. Therefore, its discriminant (from the quadratic formula) must be nonpositive. In our case, this means

$$\text{Discriminant} = 4|\langle X, Y \rangle|^2 - 4\|X\|^2 \|Y\|^2 \le 0.$$

Schwarz's inequality follows by rearranging this inequality.

If $\langle X, Y \rangle = \|X\| \, \|Y\|$, then the preceding discriminant is zero, which means that the equation $\|X - tY\|^2 = 0$ has a double real root, $\hat{t}$. In particular, $X - \hat{t}Y = 0$ or $X = \hat{t}Y$, which implies that $\langle X, Y \rangle = \hat{t}\|Y\|^2$. On the other hand, $\langle X, Y \rangle = \|X\| \, \|Y\|$ is nonnegative and therefore $\hat{t} \ge 0$. Thus $X = \hat{t}Y$ is a nonnegative multiple of Y, as claimed. The converse (i.e., if X is a nonnegative multiple of Y, then $\langle X, Y \rangle = \|X\| \, \|Y\|$) is easy and left to the reader.

Proof for a Complex Inner Product Space

If V is a complex inner product space, the proof is similar. We let ϕ be an argument of $\langle X, Y \rangle$, which means

$$\langle X, Y \rangle = |\langle X, Y \rangle| e^{i\phi}.$$

Then we consider the following inequality:

$$
\begin{aligned}
0 \le \|e^{-i\phi}X - tY\|^2 &= \langle e^{-i\phi}X - tY, e^{-i\phi}X - tY \rangle \\
&= \|X\|^2 - t\left(\langle e^{-i\phi}X, Y \rangle + \langle Y, e^{-i\phi}X \rangle \right) + t^2 \|Y\|^2 \\
&= \|X\|^2 - t\left(\langle e^{-i\phi}X, Y \rangle + \overline{\langle e^{-i\phi}X, Y \rangle} \right) + t^2 \|Y\|^2 \\
&= \|X\|^2 - 2\,\text{Re}\left\{ te^{-i\phi}\langle X, Y \rangle \right\} + t^2 \|Y\|^2
\end{aligned}
$$

where "Re" stands for "the real part"; that is, if $z = x + iy$, then

$$\text{Re } z = x = \frac{z + \bar{z}}{2}.$$

In view of the choice of ϕ, the middle term is just $-2t|\langle X, Y \rangle|$ and so the term on the right equals the expression on the right side of (0.2). The rest of the argument is now the same as the argument given for the case of a real inner product space.

Proof of the Triangle Inequality

The proof of the triangle inequality now follows from the Schwarz inequality:

$$\begin{aligned}
\|X + Y\|^2 &= \langle X + Y, X + Y \rangle \\
&= \|X\|^2 + 2\operatorname{Re}\{\langle X, Y \rangle\} + \|Y\|^2 \\
&\leq \|X\|^2 + 2\|X\|\,\|Y\| + \|Y\|^2 \quad \text{by Schwarz} \\
&= (\|X\| + \|Y\|)^2.
\end{aligned}$$

Taking square roots of both sides of this inequality establishes the triangle inequality.

If the preceding inequality becomes an equality, then $\langle X, Y \rangle = \|X\|\,\|Y\|$ and the first part of the theorem implies that either X or Y is a nonnegative multiple of the other, as claimed.   ♦

0.5   Orthogonality

0.5.1   Definitions and Examples

For the standard inner product in R^3, the law of cosines is

$$\langle X, Y \rangle = \|X\|\,\|Y\| \cos(\theta), \quad \theta = \text{angle between } X \text{ and } Y,$$

which implies that X and Y are orthogonal (perpendicular) if and only if $\langle X, Y \rangle = 0$. We shall make this equation the definition of orthogonality in general.

DEFINITION 0.12   *Suppose V is an inner product space.*

- *The vectors X and Y in V are said to be* orthogonal *if $\langle X, Y \rangle = 0$.*

- *The collection of vectors e_i, $i = 1, \ldots, N$, is said to be orthonormal if each e_i has unit length, $\|e_i\| = 1$, and e_i and e_j are orthogonal for $i \neq j$.*

- *Two subspaces V_1 and V_2 of V are said to be orthogonal if each vector in V_1 is orthogonal to every vector in V_2.*

An *orthonormal basis* or *orthonormal system* for V is a basis of vectors for V that is orthonormal.

EXAMPLE 0.13

The line $y = x$ generated by the vector $(1, 1)$ is orthogonal to the line $y = -x$ generated by $(1, -1)$.   ■

EXAMPLE 0.14

The line $x/2 = -y = z/3$ in R^3, which points in the direction of the vector $(2, -1, 3)$, is orthogonal to the plane $2x - y + 3z = 0$. ∎

EXAMPLE 0.15

For the space $L^2([0, 1])$, any two functions where the first function is zero on the set where the second is nonzero will be orthogonal. For example, if $f(t)$ is nonzero only on the interval $0 \leq t < 1/2$ and $g(t)$ is nonzero only on the interval $1/2 \leq t < 1$, then $f(t)\overline{g(t)}$ is always zero. Therefore, $\langle f, g \rangle = \int_0^1 f(t)\overline{g(t)}\, dt = 0$. ∎

EXAMPLE 0.16

Let

$$\phi(t) = \begin{cases} 1, & \text{if } 0 \leq t < 1 \\ 0, & \text{otherwise} \end{cases} \qquad \psi(t) = \begin{cases} 1, & \text{if } 0 \leq t < 1/2 \\ -1, & \text{if } 1/2 \leq t < 1 \\ 0, & \text{otherwise.} \end{cases}$$

Then ϕ and ψ are orthogonal in $L^2[0, 1]$ because

$$\langle \phi, \psi \rangle = \int_0^{1/2} 1\, dt - \int_{1/2}^1 1\, dt = 0.$$

In contrast to the previous example, note that ϕ and ψ are orthogonal and yet ϕ and ψ are nonzero on the same set, namely the interval $0 \leq t \leq 1$. The function ϕ is called the *scaling function* and the function ψ is called the *wavelet function* for the Haar system. We shall revisit these functions in the later chapters on wavelets. ∎

EXAMPLE 0.17

The function $f(t) = \sin t$ and $g(t) = \cos t$ are orthogonal in $L^2([-\pi, \pi])$, because

$$\langle f, g \rangle = \int_{-\pi}^{\pi} \sin(t) \cos(t)\, dt$$

$$= \frac{1}{2} \int_{-\pi}^{\pi} \sin(2t)\, dt$$

$$= \frac{-1}{4} \cos(2t) \Big|_{-\pi}^{\pi}$$

$$= 0.$$

Since $\int_{-\pi}^{\pi} \sin^2(t)\, dt = \int_{-\pi}^{\pi} \cos^2(t)\, dt = \pi$, the functions $\frac{\sin(t)}{\sqrt{\pi}}$ and $\frac{\cos(t)}{\sqrt{\pi}}$ are orthonormal in $L^2([-\pi, \pi])$. More generally, we shall show in Chapter 1 that the functions

$$\frac{\cos nt}{\sqrt{\pi}}, \frac{\sin(nt)}{\sqrt{\pi}}, \quad n = 1, 2, \dots$$

are orthonormal. This fact will be very important in our development of Fourier series.   ∎

Vectors can be expanded easily in terms of an orthonormal basis, as the following theorem shows.

THEOREM 0.18 *Suppose V_0 is a subspace of an inner product space V. Suppose $\{e_1, \ldots e_N\}$ is an orthonormal basis for V_0. If $v \in V_0$, then*

$$v = \sum_{j=1}^{N} \langle v, e_j \rangle e_j.$$

Proof Since $\{e_1, \ldots, e_N\}$ is a basis for V_0, any vector $v \in V_0$ can be uniquely expressed as a linear combination of the e_j:

$$v = \sum_{j=1}^{N} \alpha_j e_j.$$

To evaluate the constant α_k, take the inner product of both sides with e_k:

$$\langle v, e_k \rangle = \sum_{j=1}^{N} \langle \alpha_j e_j, e_k \rangle.$$

The only nonzero term on the right occurs when $j = k$ since the e_j are orthonormal. Therefore,

$$\langle v, e_k \rangle = \alpha_k \langle e_k, e_k \rangle = \alpha_k.$$

Thus, $\alpha_k = \langle v, e_k \rangle$, as desired.   ♦

0.5.2   Orthogonal Projections

Suppose $\{e_1, \ldots e_N\}$ is an orthonormal collection of vectors in an inner product space V. If v lies in the span of $\{e_1, \ldots e_N\}$, then, as Theorem 0.18 demonstrates, the equation

$$v = \sum_{j=1}^{N} \alpha_j e_j \tag{0.3}$$

is satisfied with $\alpha_j = \langle v, e_j \rangle$. If v does not lie in the linear span of $\{e_1, \ldots e_N\}$, then solving (0.3) for α_j is impossible. In this case, the best we can do is to determine the vector v_0 belonging to the linear span of $\{e_1, \ldots e_N\}$ that comes as close as possible to v. More generally, suppose V_0 is a subspace of the inner product space V and suppose $v \in V$ is a vector that is *not* in V_0 (see Figure 5). How can we determine the vector $v_0 \in V_0$ that is closest to v? This vector (v_0) has a special name in the following definition.

DEFINITION 0.19   *Suppose V_0 is a finite dimensional subspace of an inner product space V. For any vector $v \in V$, the orthogonal projection of v onto V_0 is the unique vector $v_0 \in V_0$ that is closest to v; that is,*

$$\|v - v_0\| = \min_{w \in V_0} \|v - w\|.$$

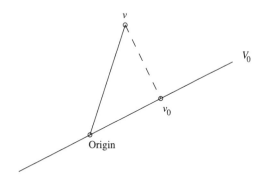

Figure 5 Orthogonal projection of v onto V_0

As Figure 5 indicates, the vector v_0, which is closest to v, must be chosen so that $v - v_0$ (the vector from v_0 to v) is orthogonal to V_0. Of course, figures are easily drawn when the underlying vector space is R^2 or R^3. In a more complicated inner product space, such as L^2, figures are an abstraction, which may or may not be accurate (e.g., an element in L^2 is not really an a point in the plane as in Figure 5). The following theorem states that our intuition in R^2 regarding orthogonality is accurate in a general inner product space.

THEOREM 0.20   *Suppose V_0 is a finite dimensional subspace of an inner product space V. Let v be any element in V. Then its orthogonal projection, v_0, has the following property: $v - v_0$ is orthogonal to every vector in V_0.*

Proof   We first show that if v_0 is the closest vector to v, then $v - v_0$ is orthogonal to every vector $w \in V_0$. Consider the function

$$f(t) = \|v_0 + tw - v\|^2 \quad t \in R,$$

which describes the square of the distance between $v_0 + tw \in V_0$ and v. If v_0 is the closest element of V_0 to v, then f is minimized when $t = 0$. For simplicity, we will consider the case where the underlying inner product space V is real. Expanding f, we have

$$f(t) = \langle (v_0 - v) + tw, \ (v_0 - v) + tw \rangle$$
$$= \|v_0 - v\|^2 + 2t\langle v_0 - v, w \rangle + t^2 \|w\|^2.$$

Since f is minimized when $t = 0$, its derivative at $t = 0$ must be zero. We have

$$f'(t) = 2\langle v_0 - v, w \rangle + 2t\|w\|^2.$$

So

$$0 = f'(0) = 2\langle v_0 - v, w \rangle \tag{0.4}$$

and we conclude that $v_0 - v$ is orthogonal to w.

The converse also holds: If $v_0 - v$ is orthogonal to w, then from (0.4), $f'(0) = 0$. Since $f(t)$ is a nonnegative quadratic polynomial in t, this critical point $t = 0$ must correspond to a minimum. Therefore, $\|v_0 + tw - v\|^2$ is minimized when $t = 0$. Since w is an arbitrarily chosen vector in V_0, we conclude that v_0 is the closest vector in V_0 to v.   ◆

In terms of an orthonormal basis for V_0, the projection of a vector v onto V_0 is easy to compute, as the following theorem states.

THEOREM 0.21 *Suppose V is an inner product space and V_0 is an N-dimensional subspace with orthonormal basis $\{e_1, e_2, \ldots e_N\}$. The orthogonal projection of a vector $v \in V$ onto V_0 is given by*

$$v_0 = \sum_{j=1}^{N} \alpha_j e_j \quad \text{with} \quad \alpha_j = \langle v, e_j \rangle.$$

Note. In the special case that v belongs to V_0, then v equals its orthogonal projection, v_0. In this case, the preceding formula for $v_0 = v$ is the same as the one given in Theorem 0.18.

Proof Let $v_0 = \sum_{j=1}^{N} \alpha_j e_j$ with $\alpha_j = \langle v, e_j \rangle$. In view of Theorem 0.20, we must show that $v - v_0$ is orthogonal to any vector $w \in V_0$. Since $e_1, \ldots e_N$ is a basis for V_0, it suffices to show that $v - v_0$ is orthogonal to each e_k, $k = 1, \ldots, N$. We have

$$\langle v - v_0, e_k \rangle = \langle v - \sum_{j=1}^{N} \alpha_j e_j, \ e_k \rangle.$$

Since $e_1, \ldots, e_N$ are orthonormal, the only contributing term to the sum is when $j = k$:

$$\begin{aligned}
\langle v - v_0, e_k \rangle &= \langle v, e_k \rangle - \alpha_k \langle e_k, e_k \rangle \\
&= \langle v, e_k \rangle - \alpha_k \quad \text{since } \|e_k\| = 1 \\
&= 0 \quad \text{since } \alpha_k = \langle v, e_k \rangle.
\end{aligned}$$

Thus, $v - v_0$ is orthogonal to each e_k and hence to all of V_0, as desired.   ◆

EXAMPLE 0.22

Let V_0 be the space spanned by $\cos x$ and $\sin x$ in $L^2([-\pi, \pi])$. As computed in Example 0.17, the functions

$$e_1 = \frac{\cos x}{\sqrt{\pi}} \quad \text{and} \quad e_2 = \frac{\sin x}{\sqrt{\pi}}$$

are orthonormal in $L^2([-\pi, \pi])$. Let $f(x) = x$. The projection of f onto V_0 is given by

$$f_0 = \langle f, e_1 \rangle e_1 + \langle f, e_2 \rangle e_2.$$

Now, $f(x) \cos(x) = x \cos(x)$ is odd and so $\langle f, e_1 \rangle = \frac{1}{\sqrt{\pi}} \int_{-\pi}^{\pi} f(x) \cos(x) \, dx = 0$. For the other term,

$$\langle f, e_2 \rangle = \frac{1}{\sqrt{\pi}} \int_{-\pi}^{\pi} x \sin(x) \, dx$$

$$= 2\sqrt{\pi} \quad \text{[integrate by parts.]}$$

Therefore, the projection of $f(x) = x$ onto V_0 is given by

$$f_0 = \langle f, e_2 \rangle e_2 = 2\sqrt{\pi} \, \frac{\sin x}{\sqrt{\pi}} = 2 \sin(x). \qquad \blacksquare$$

EXAMPLE 0.23

Consider the space V_1 that is spanned by $\phi(x) = 1$ on $0 \leq x < 1$ and

$$\psi(x) = \begin{cases} 1, & 0 \leq x < 1/2 \\ -1, & 1/2 \leq x < 1. \end{cases}$$

The functions ϕ and ψ are the Haar scaling function and wavelet function mentioned earlier. These two functions are orthonormal in $L^2([0, 1])$. Let $f(x) = x$. As you can check,

$$\langle f, \phi \rangle = \int_0^1 x \, dx = 1/2$$

and

$$\langle f, \psi \rangle = \int_0^{1/2} x \, dx - \int_{1/2}^1 x \, dx = -1/4.$$

So the orthogonal projection of the function f onto V_1 is given by

$$f_0 = \langle f, \phi \rangle \phi + \langle f, \psi \rangle \psi = \phi/2 - \psi/4 = \begin{cases} 1/4, & 0 \leq x < 1/2 \\ 3/4, & 1/2 \leq x < 1. \end{cases} \qquad \blacksquare$$

The set of vectors that are orthogonal to a given subspace has a special name.

DEFINITION 0.24 *Suppose V_0 is a subspace of an inner product space V. The orthogonal complement of V_0, denoted $V_0^\perp$, is the set of all vectors in V that are orthogonal to V_0; that is,*

$$V_0^\perp = \{v \in V; \ \langle v, w \rangle = 0 \text{ for every } w \in V_0\}.$$

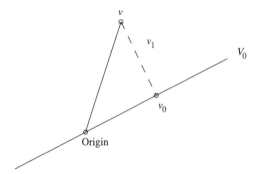

Figure 6 $V = V_0 \oplus V_0^\perp$; $v = v_0 + v_1$ with $v_0 \in V_0$, $v_1 \in V_0^\perp$

As Figure 6 indicates, each vector can be written as a sum of a vector in V_0 and a vector in $V_0^\perp$. The intuition from this Euclidean figure is accurate for more general inner product spaces, as the following theorem demonstrates.

THEOREM 0.25   *Suppose V_0 is a finite dimensional subspace of an inner product space V. Each vector $v \in V$ can be written uniquely as $v = v_0 + v_1$, where v_0 belongs to V_0 and v_1 belongs to $V_0^\perp$; that is,*

$$V = V_0 \oplus V_0^\perp.$$

Proof   Suppose v belongs to V and let v_0 be its orthogonal projection onto V_0. Let $v_1 = v - v_0$; then

$$v = v_0 + (v - v_0) = v_0 + v_1.$$

By Theorem 0.20, v_1 is orthogonal to every vector in V_0. Therefore, v_1 belongs to $V_0^\perp$.   ♦

EXAMPLE 0.26
_____

Consider the plane $V_0 = \{2x - y + 3z = 0\}$. The set of vectors

$$\left\{ e_1 = \frac{1}{\sqrt{21}}(1, -4, -2) \quad \text{and} \quad e_2 = \frac{1}{\sqrt{6}}(2, 1, -1) \right\}$$

forms an orthonormal basis for V_0. So given $v = (x, y, z) \in R^3$, the vector

$$v_0 = \langle v, e_1 \rangle e_1 + \langle v, e_2 \rangle e_2$$
$$= \left(\frac{x - 4y - 2z}{21} \right)(1, -4, -2) + \left(\frac{2x + y - z}{6} \right)(2, 1, -1)$$

is the orthogonal projection of v onto the plane V_0.

The vector $e_3 = (2, -1, 3)/\sqrt{14}$ is a unit vector that is perpendicular to this plane. So

$$v_1 = \langle v, e_3 \rangle e_3 = \frac{(2x - y + 3z)}{14}(2, -1, 3)$$

is the orthogonal projection of $v = (x, y, z)$ onto $V_0^\perp$.   ■

The theorems in this section are valid for certain infinite dimensional subspaces, but a discussion of infinite dimensions involves more advanced ideas from functional analysis (see [18]).

0.5.3 Gram-Schmidt Orthogonalization

Theorems 0.18 and 0.21 indicate the importance of finding an orthonormal basis. Without an orthonormal basis, the computation of an orthogonal projection onto a subspace is much more difficult. If an orthonormal basis is not readily available, then the Gram-Schmidt orthogonalization procedure describes a way to construct one for a given subspace.

THEOREM 0.27 *Suppose V_0 is a subspace of dimension N of an inner product space V. Let v_j, $j = 1, \ldots, N$ be a basis for V_0. Then there is an orthonormal basis $\{e_1, \ldots, e_N\}$ for V_0 such that each e_j is a linear combination of $v_1, \ldots, v_j$.*

Proof We first define $e_1 = v_1/\|v_1\|$. Clearly, e_1 has unit length. Let v_0 be the orthogonal projection of v_2 onto the line spanned by e_1. From Theorem 0.21,

$$v_0 = \langle v_2, e_1 \rangle e_1.$$

Figure 7 suggests that the vector from v_0 to v_2 is orthogonal to e_1. The vector E_2 is

$$\begin{aligned} E_2 &= v_2 - v_0 \quad \text{(the vector from } v_0 \text{ to } v_2) \\ &= v_2 - \langle v_2, e_1 \rangle e_1 \end{aligned}$$

and

$$\langle E_2, e_1 \rangle = \langle v_2 - \langle v_2, e_1 \rangle e_1, e_1 \rangle = \langle v_2, e_1 \rangle - \langle v_2, e_1 \rangle \langle e_1, e_1 \rangle = 0,$$

which confirms our intuition from Figure 7.

Note that E_2 cannot equal zero for otherwise v_2 and e_1 (and hence v_2 and v_1) would be linearly dependent. To get a vector of unit length, we define $e_2 = E_2/\|E_2\|$. Both e_1 and e_2 are orthogonal to each other, and since e_1 is a multiple of v_1, the vector e_2 is a linear combination of v_1 and v_2.

If $N > 2$, then we continue the process. We consider the orthogonal projection of v_3 onto the space spanned by e_1 and e_2:

$$\text{Orthogonal projection} = v_0 = \langle v_3, e_1 \rangle e_1 + \langle v_3, e_2 \rangle e_2.$$

Then let

$$E_3 = v_3 - v_0 = v_3 - (\langle v_3, e_1 \rangle e_1 + \langle v_3, e_2 \rangle e_2)$$

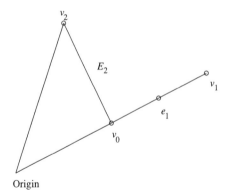

Figure 7 Gram-Schmidt orthogonalization

and set $e_3 = E_3/\|E_3\|$. The same argument as before (for E_2) shows that E_3 is orthogonal to both e_1 and e_2. Thus, $\{e_1, e_2, e_3\}$ is an orthonormal set of vectors that span the same space as v_1, v_2 and v_3. The pattern is now clear.   ♦

0.6   Linear Operators and Their Adjoints

0.6.1   Linear Operators

First, we recall the definition of a linear map.

DEFINITION 0.28   *A linear operator (or map) between a vector space V and a vector space W is a function $T : V \to W$ which satisfies*

$$T(\alpha v + \beta w) = \alpha T(v) + \beta T(w) \quad for \ v, \ w \in V \quad \alpha, \beta \in C.$$

If V and W are finite dimensional, then T is often identified with its matrix representation with respect to a given choice of bases, say $\{v_1, \ldots, v_n\}$ for V and $\{w_1 \ldots w_m\}$ for W. For each $1 \leq j \leq n$, Tv_j belongs to W and therefore it can be expanded in terms of $w_1, \ldots, w_m$:

$$T(v_j) = \sum_{i=1}^{m} a_{ij} w_i \tag{0.5}$$

where a_{ij} are complex numbers. The value of $T(v)$ for any vector $v = \sum_j x_j v_j \in V$ (with $x_j \in C$) can be computed by

$$T(v) = T\left(\sum_{j=1}^{n} x_j v_j\right) = \sum_{j=1}^{n} x_j T(v_j) = \sum_{i=1}^{m} \sum_{j=1}^{n} (a_{ij} x_j) w_i \quad \text{from (0.5)}$$

$$= \sum_{i=1}^{m} c_i w_i.$$

The coefficient of w_i is $c_i = \sum_{j=1}^{n} a_{ij} x_j$, which can be identified with the ith entry in the following matrix product:

$$\begin{pmatrix} a_{11} & \cdots & a_{1n} \\ \vdots & \ddots & \vdots \\ a_{m1} & \cdots & a_{mn} \end{pmatrix} \begin{pmatrix} x_1 \\ \vdots \\ x_n \end{pmatrix}.$$

Thus, the matrix (a_{ij}) is determined by how the basis vectors $v_j \in V$ are mapped into the basis vectors w_i of W [see Equation (0.5)]. The matrix then determines how an arbitrary vector v maps into W.

In words, a linear operator $T : V \to W$ is said to be *bounded* if it maps the unit ball in V to a bounded set in W. This means that there is a number $0 \leq M < \infty$ such that

$$\{Tv; \ v \in V \text{ with } \|v\| \leq 1 \ \} \subset \{w \in W; \ \|w\| \leq M\}.$$

In this case, the *norm* of T is defined to be the smallest such M. All linear maps between finite dimensional inner product spaces are bounded. As another example, the orthogonal projection map from any inner product space to any of its subspaces (finite or infinite dimensions) is a bounded linear operator.

0.6.2 Adjoints

If V and W are inner product spaces, then we will sometimes need to compute $\langle T(v), w \rangle_W$ by shifting the operator T to the other side of the inner product. In other words, we want to write

$$\langle T(v), w \rangle_W = \langle v, T^*(w) \rangle_V$$

for some operator $T^* : W \to V$. We formalize the definition of such a map as follows.

DEFINITION 0.29 *If $T : V \to W$ is a linear operator between two inner product spaces, the* adjoint *of T is the linear operator $T^* : W \to V$ that satisfies*

$$\langle T(v), w \rangle_W = \langle v, T^*(w) \rangle_V.$$

Every bounded linear operator between two inner product spaces always has an adjoint. Here are two examples of adjoints of linear operators.

EXAMPLE 0.30

Let $V = C^n$ and $W = C^m$ with the standard inner products. Suppose $T : C^n \to C^m$ is a linear operator with matrix $a_{ij} \in C$ with respect to the standard basis

$$e_j = (0, \ldots, 1, \ldots 0) \quad (1 \text{ in the } j\text{th component}).$$

If $X = (x_1, \ldots, x_n) \in C^n$ and $Y = (y_1, \ldots, y_m) \in C^m$, then

$$\langle T(X), Y \rangle = \sum_{i=1}^{m} \sum_{j=1}^{n} a_{ij} x_j \overline{y_i}$$

$$= \sum_{j=1}^{n} x_j \left(\overline{\sum_{i=1}^{m} \overline{a_{ij}} y_i} \right) \quad \text{(switch order of summation)}$$

$$= \sum_{j=1}^{n} x_j \left(\overline{\sum_{i=1}^{m} a_{ji}^* y_i} \right)$$

where $a_{ji}^* = \overline{a_{ij}}$ (the conjugate of the transpose). The right side is $\langle X, T^*(Y) \rangle$, where the jth component of $T^*(Y)$ is $\sum_{i=1}^{m} a_{ji}^* y_i$. *Thus the matrix for the adjoint of T is the conjugate of the transpose of the matrix for T.* ∎

EXAMPLE 0.31

Suppose g is a bounded function on the interval $a \leq x \leq b$. Let $T_g : L^2([a,b]) \to L^2([a,b])$ be defined by

$$T_g(f)(x) = g(x)f(x).$$

The adjoint of T_g is just

$$T_g^*(h)(x) = \overline{g(x)}h(x)$$

because

$$\langle T_g(f), h \rangle = \int_a^b g(x)f(x)\overline{h(x)}\, dx = \int_a^b f(x)\overline{\overline{g(x)}h(x)}\, dx = \langle f, \overline{g}h \rangle. \quad ∎$$

The next theorem computes the adjoint of the composition of two operators.

THEOREM 0.32 *Suppose $T_1 : V \to W$ and $T_2 : W \to U$ are bounded linear operators between inner product spaces. Then $(T_2 \circ T_1)^* = T_1^* \circ T_2^*$.*

Proof For $v \in V$ and $u \in U$, we have

$$\langle T_2(T_1(v)), u \rangle = \langle T_1(v), T_2^*(u) \rangle = \langle v, T_1^*(T_2^*(u)) \rangle.$$

On the other hand, the definition of *adjoint* implies

$$\langle T_2(T_1(v)), u \rangle = \langle v, (T_2 \circ T_1)^* u \rangle. \tag{0.6}$$

Therefore,

$$\langle v, (T_2 \circ T_1)^* u \rangle = \langle v, T_1^*(T_2^*(u)) \rangle$$

for all $v \in V$. By Exercise 17, if u_0 and u_1 are vectors in V with $\langle v, u_0 \rangle = \langle v, u_1 \rangle$ for all $v \in V$, then $u_0 = u_1$. Therefore, from (0.6), we conclude

$$(T_2 \circ T_1)^* u = T_1^*(T_2^*(u))$$

as desired. ◆

In the next theorem, we compute the adjoint of an orthogonal projection.

THEOREM 0.33 *Suppose V_0 is a subspace of an inner product space V. Let $\pi : V \to V_0$ be the map that assigns to each $v \in V$ its orthogonal projection onto V_0. Then the adjoint of π is the inclusion map $\iota : V_0 \to V$, where for $v_0 \in V_0$, $\iota(v_0) = v_0$ thought of as an element of V.*

Proof From Theorem 0.25, each vector $v \in V$ can be written as $v = v_0 + v_1$, where $v_0 \in V_0$ and v_1 is orthogonal to V_0. Note that $\pi(v) = v_0$. Therefore, for $u_0 \in V_0$,

$$\langle \pi(v), u_0 \rangle_{V_0} = \langle v_0, u_0 \rangle_{V_0}.$$

Now the subspace V_0 inherits its inner product from V. Thus

$$\begin{aligned}
\langle \pi(v), u_0 \rangle_{V_0} &= \langle v_0, u_0 \rangle_V \\
&= \langle v_0 + v_1, u_0 \rangle_V \quad \text{since } v_1 \in V_0^\perp, \text{ and } u_0 \in V_0 \\
&= \langle v, u_0 \rangle_V.
\end{aligned}$$

So

$$\langle \pi(v), u_0 \rangle_{V_0} = \langle v, u_0 \rangle_V.$$

In addition,

$$\langle \pi(v), u_0 \rangle_{V_0} = \langle v, \pi^*(u_0) \rangle_V$$

by the definition of *adjoint*. By comparing the two expressions for $\langle \pi(v), u_0 \rangle_{V_0}$, we conclude

$$\langle v, u_0 \rangle_V = \langle v, \pi^*(u_0) \rangle_V$$

for all $v \in V$. By Exercise 17, $\pi^*(u_0) = u_0$, as claimed. ♦

0.7 Least Squares and Linear Predictive Coding

In this section, we apply some of the basic facts about linear algebra and inner product spaces to the topic of least squares analysis. As motivation, we first describe how to find the best fit line to a collection of data. Then we present the general idea behind the least squares algorithm. As a further application of least squares, we present the ideas behind linear predictive coding, which is a data compression algorithm. In the next chapter, we use least squares to develop a procedure for approximating signals (or functions) by trigonometric polynomials.

0.7.1 Best Fit Line for Data

Consider the following problem. Suppose data points x_i and y_i for $i = 1, \ldots N$ are given with $x_i \neq x_j$ for $i \neq j$. We wish to find the equation of the line $y = mx + b$ that comes closest to fitting all the data. Figure 8 gives an example of four data points (indicated by the small circles) as well as the graph of the line that comes closest to passing through these four points.

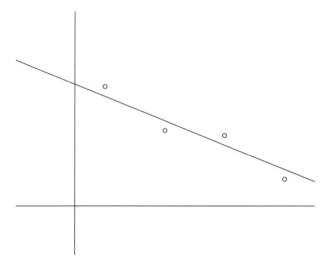

Figure 8 Least squares approximation

The word *closest* here means that the sum of the squares of the errors (between the data and the line) is smaller than the corresponding error with any other line. Suppose the line we seek is $y = mx + b$. As Figure 9 demonstrates, the error between this line at $x = x_i$ and the data point (x_i, y_i) is $|y_i - (mx_i + b)|$.

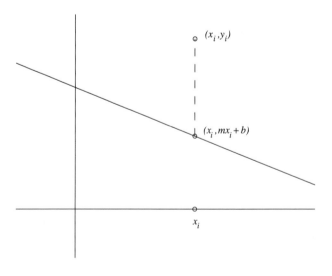

Figure 9 Error at x_i is $|y_i - (mx_i + b)|$ (length of dashed line)

Therefore, we seek numbers m and b that minimize the quantity

$$E = \sum_{i=1}^{N} |y_i - (mx_i + b)|^2.$$

The quantity E can be viewed as the square of the distance (in R^N) from the vector

$$Y = \begin{pmatrix} y_1 \\ y_2 \\ \vdots \\ y_N \end{pmatrix}$$

and the vector $mX + bU$, where

$$X = \begin{pmatrix} x_1 \\ x_2 \\ \vdots \\ x_N \end{pmatrix} \quad \text{and} \quad U = \begin{pmatrix} 1 \\ 1 \\ \vdots \\ 1 \end{pmatrix}. \tag{0.7}$$

As m and b vary over all possible real numbers, the expression $mX + bU$ sweeps out a two-dimensional plane, M in R^N. Thus our problem of least squares has the following geometric interpretation: Find the point $P = mX + bU$ on M that is closest to the point Y (see Figure 10).

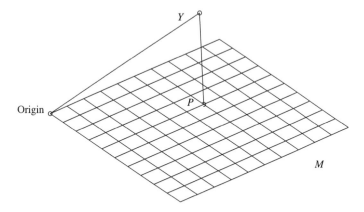

Figure 10 P is the closest point on the plane M to the point Y

The point P must be the orthogonal projection of Y onto M. In particular, $Y - P$ must be orthogonal to M. Since M is generated by the vectors X and U, $Y - P$ must be orthogonal to both X and U. Therefore, we seek the point $P = mX + bU$ that satisfies the following two equations:

$$0 = \langle (Y - P), X \rangle = \langle (Y - (mX + bU)), X \rangle$$
$$0 = \langle (Y - P), U \rangle = \langle (Y - (mX + bU)), U \rangle$$

or

$$\langle X, Y \rangle = m \langle X, X \rangle + b \langle X, U \rangle$$
$$\langle U, Y \rangle = m \langle X, U \rangle + b \langle U, U \rangle.$$

These equations can be rewritten in matrix form as

$$\begin{pmatrix} x_1 & \cdots & x_N \\ 1 & \cdots & 1 \end{pmatrix} \begin{pmatrix} y_1 \\ \vdots \\ y_N \end{pmatrix} = \begin{pmatrix} x_1 & \cdots & x_N \\ 1 & \cdots & 1 \end{pmatrix} \begin{pmatrix} x_1 & 1 \\ \vdots & \vdots \\ x_N & 1 \end{pmatrix} \begin{pmatrix} m \\ b \end{pmatrix}.$$

Keep in mind that the x_i and y_i are the known data points. The solution to our least squares problem is obtained by solving the preceding linear system for the unknowns m and b. This discussion is summarized in the following theorem.

THEOREM 0.34   *Suppose $X = \{x_1, x_2, \ldots, x_N\}$ and $Y = \{y_1, y_2, \ldots, y_N\}$ are two sets of data points. The equation of the line $y = mx + b$ that most closely approximates the data $(x_1, y_1), \ldots, (x_N, y_N)$ in the sense of least squares is obtained by solving the linear equation*

$$Z^T Y = Z^T Z \begin{pmatrix} m \\ b \end{pmatrix}$$

for m and b, where

$$Z = \begin{pmatrix} x_1 & 1 \\ \vdots & \vdots \\ x_N & 1 \end{pmatrix}.$$

If the x_i are distinct, then this system of equations has the following unique solution for m and b:

$$m = \frac{\langle Y, X \rangle - N \overline{x} \, \overline{y}}{\sigma_x} \quad and \quad b = \frac{\overline{y}(\sum_i x_i^2) - \overline{x} \langle X, Y \rangle}{\sigma_x}$$

where $\sigma_x = \sum_i (x_i - \overline{x})^2$, $\overline{x} = (\sum_i x_i)/N$ and $\overline{y} = (\sum_i y_i)/N$.

Proof   We leave the computation of the formulas for m and b as an exercise (see Exercise 24). The statement regarding the unique solution for m and b will follow once we show that the matrix $Z^T Z$ is nonsingular (i.e., invertible). To this end, suppose the x_i are not all the same; then the vectors X and U are linearly independent and so the matrix Z has rank 2. In addition, for any $V \in R^2$

$$\langle (Z^T Z) V, V \rangle = \langle (Z) V, (Z) V \rangle \quad \text{(see Example 0.30)}$$
$$= |(Z) V|^2$$
$$\geq 0.$$

Since Z is a matrix of maximal rank (i.e., rank 2), the only way $(Z)V$ can be zero is if V is zero. Therefore, $\langle(Z^T Z)V, V\rangle > 0$ for all nonzero V, which means that $Z^T Z$ is positive definite. In addition, the matrix $Z^T Z$ is symmetric because its transpose, $(Z^T Z)^T$, equals itself. Using a standard fact from linear algebra, this positive definite symmetric matrix must be nonsingular. Thus, the equation

$$(Z^T Z)\binom{m}{b} = Z^T(Y)$$

has a unique solution for m and b. ◆

0.7.2 General Least Squares Algorithm

Suppose Z is a $N \times q$ matrix (with possibly complex entries) and let Y be a vector in R^N (or C^N). Linear algebra is, in part, the study of the equation $ZV = Y$, which when written out in detail is

$$\begin{pmatrix} z_{11} & \cdots & z_{1q} \\ \vdots & \ddots & \vdots \\ z_{N1} & \cdots & z_{Nq} \end{pmatrix} \begin{pmatrix} v_1 \\ \vdots \\ v_q \end{pmatrix} = \begin{pmatrix} y_1 \\ \vdots \\ y_N \end{pmatrix}.$$

If $N > q$, then the equation $ZV = Y$ does not usually have a solution for $V \in C^q$ because there are more equations (N) than there are unknowns ($v_1, \ldots, v_q$). If there is no solution, the problem of least squares asks for the next best quantity: Find the vector $V \in C^q$ such that ZV is as close as possible to Y.

In the case of finding the best fit line to a set of data points (x_i, y_i), $i = 1 \ldots N$, the matrix Z is

$$Z = \begin{pmatrix} x_1 & 1 \\ \vdots & \vdots \\ x_N & 1 \end{pmatrix} \tag{0.8}$$

and the vectors Y and V are

$$Y = \begin{pmatrix} y_1 \\ \vdots \\ y_N \end{pmatrix} \quad \text{and} \quad V = \binom{m}{b}.$$

In this case, the matrix product ZV is

$$\begin{pmatrix} x_1 & 1 \\ \vdots & \vdots \\ x_N & 1 \end{pmatrix} \binom{m}{b} = \begin{pmatrix} mx_1 + b \\ \vdots \\ mx_N + b \end{pmatrix} = mX + bU$$

where X and U are the vectors given in (0.7). Thus, finding the $V = (m, b)$ so that ZV is closest to Y is equivalent to finding the slope and y-intercept of the best fit line to the data (x_i, y_i), $i = 1 \ldots N$, as in the last section.

The solution to the general least squares problem is given in the following theorem.

THEOREM 0.35   *Suppose Z is a $N \times q$ matrix (with possibly complex entries) of maximal rank and with $N \geq q$. Let Y be a vector in R^N (or C^N). There is a unique vector $V \in C^q$ such that ZV is closest to Y. Moreover, the vector V is the unique solution to the matrix equation*

$$Z^* Y = Z^* Z V.$$

If Z is a matrix with real entries, then the preceding equation becomes

$$Z^T Y = Z^T Z V.$$

Note that in the case of the best fit line, the matrix Z in (0.8) and the equation $Z^T Y = Z^T Z V$ are the same as those given in Theorem 0.34.

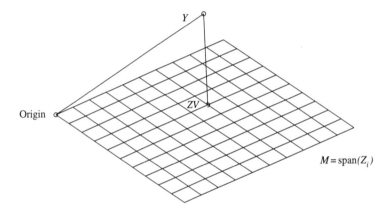

Figure 11 $Y - ZV$ must be orthogonal to $M = \text{span}\{Z_1, \ldots Z_q\}$

Proof   The proof of this theorem is similar to the proof given in the construction of the best fit line. We let $Z_1, \ldots, Z_q$ be the columns of the matrix Z. Then $ZV = v_1 Z_1 + \cdots + v_q Z_q$ is a point that lies in the subspace $M \subset C^N$ generated by $Z_1, \ldots, Z_q$. We wish to find the point ZV that is closest to Y. As in Figure 11, $Y - ZV$ must be orthogonal to M, or, equivalently, $Y - ZV$ must be orthogonal to $Z_1, \ldots, Z_q$ that generate M. Thus

$$\langle Y - ZV, Z_i \rangle = 0 \quad 1 \leq i \leq q.$$

These equations can be written succinctly as

$$Z^*(Y - ZV) = 0$$

because the ith component of this (vector) equation is the inner product of $Y - ZV$ with Z_i. This equation can be rearranged to read

$$Z^* Y = Z^* Z V$$

as claimed in the theorem.

The matrix Z^*Z has dimension $q \times q$ and by the same arguments used in the proof of Theorem 0.34, you can show that this matrix is nonsingular (using the fact that Z has maximal rank). Therefore, the equation

$$Z^*Y = Z^*ZV$$

has a unique solution $V \in C^q$ as claimed. ♦

EXAMPLE 0.36

Suppose a set of data points $\{(x_i, y_i), \ i = 1, \ldots N\}$ behaves in a quadratic rather than a linear fashion. Then a best fit quadratic equation of the form $y = ax^2 + bx + c$ can be found. In this case, we seek a, b, and c that minimize the quantity

$$E = \sum_{i=1}^{N} |y_i - (ax_i^2 + bx_i + c)|^2.$$

We can apply Theorem 0.35 with

$$Z = \begin{pmatrix} x_1^2 & x_1 & 1 \\ \vdots & \vdots & \vdots \\ x_N^2 & x_N & 1 \end{pmatrix} \quad V = \begin{pmatrix} a \\ b \\ c \end{pmatrix} \quad \text{and} \quad Y = \begin{pmatrix} y_1 \\ \vdots \\ y_N \end{pmatrix}.$$

From Theorem 0.35, the solution $V = (a, b, c)$ to this least squares problem is the solution to $Z^T Z V = Z^T Y$. Exercise 28 asks you to solve this system with specific numerical data. ∎

0.7.3 Linear Predictive Coding

Here, we will apply the least squares analysis procedure to the problem of efficiently transmitting a signal. As mentioned earlier, computers can process millions, and in some cases billions, of instructions per second. However, if the output must be transmitted from one location to another (say a picture downloaded from the Web), the signal must often be sent over telephone lines or some other medium that can only transmit thousands of bytes per second (in the case of telephone lines, this rate is currently about 60 kilobytes per second). Therefore, instead of transmitting all the data points of a signal, some sort of coding algorithm (data compression) is applied so that only the essential parts of the signal are transmitted.

Let us suppose we are transmitting a signal, which after some discretization process can be thought of as a long string of numbers

$$x_1, x_2, x_3, \ldots$$

(zeros and ones, perhaps). For simplicity, we will assume that each x_i is real. Often, there is a repetition of some pattern of the signal (redundancy). If the repetition is perfect, say 1,1,0, 1,1,0, 1,1,0, etc., then there would be no need to send all the digits. We would only need to transmit the pattern 1,1,0 and the number of times this pattern is repeated. Usually, however, there is not a perfect repetition of a pattern, but there may be some pattern that is nearly repetitive. For example, the rhythm of a beating heart is nearly, but not exactly, repetitive (if it is a healthy heart). If there is a near repetition of some pattern, then the following linear predictive coding procedure can achieve significant compression of data.

Main Idea. The idea behind linear predictive coding is to divide up the data into blocks of length N, where N is a large number.

$$\{x_1 \ldots x_N\}, \{x_{N+1} \ldots x_{2N}\}, \{x_{2N+1} \ldots x_{3N}\} \ldots$$

Let's consider the first block of data $x_1, \ldots, x_N$. We choose a number p that should be small compared with N. The linear predictive coding scheme will provide the best results (best compression) if p is chosen close to the number of digits in the near repetitive pattern of this block of data. Next, we try to find numbers $a_1, a_2, \ldots, a_p$ that minimize the terms

$$e(n) = x_n - \sum_{k=1}^{p} a_k x_{n-k} \quad \text{for } p+1 \le n \le N \tag{0.9}$$

in the sense of least squares. Once this is done (the details of which will be presented later), then the idea is to transmit $x_1 \ldots x_p$ as well as $a_1, \ldots, a_p$. Instead of transmitting $x_{p+1}, x_{p+2}, \ldots$, we use the following scheme starting with $n = p+1$. If $e(p+1)$ is smaller than some specified tolerance, then we can treat $e(p+1)$ as zero. By letting $n = p+1$ and $e(p+1) = 0$ in (0.9), we have

$$x_{p+1} = \sum_{k=1}^{p} a_k x_{p+1-k} = a_1 x_p + a_2 x_{p-1} + a_3 x_{p-2} + \cdots + a_p x_1.$$

There is no need to transmit x_{p+1} because the data $x_1 \ldots x_p$ as well as $a_1 \ldots a_p$ have already been transmitted and so the receiver can reconstruct x_{p+1} according to the preceding formula. If $e(p+1)$ is larger than the specified tolerance, then x_{p+1} [or, equivalently, $e(p+1)$] needs to be transmitted.

Once the receiver has reconstructed (or received) x_{p+1}, n can be incremented to $p+2$ in (0.9). If e_{p+2} is smaller than the tolerance, then x_{p+2} does not need to be transmitted and the receiver can reconstruct x_{p+2} by setting $e(p+2) = 0$ in (0.9) giving

$$x_{p+2} = a_1 x_{p+1} + \cdots + a_p x_2.$$

The rest of the $x_{p+3}, \ldots, x_N$ can be reconstructed by the receiver in a similar fashion.

The hope is that if the a_i have been chosen to minimize $\{e(p+1), \ldots, e(N)\}$ in the sense of least squares, then most of the $|e(n)|$ will be less than the specified tolerance and therefore most of the x_n can be reconstructed by the receiver and not actually transmitted. The result is that instead of transmitting N

pieces of data (i.e., $x_1, \ldots, x_N$), we only need to transmit $2p$ pieces of data (i.e., $a_1, \ldots, a_p$ and $x_1, \ldots, x_p$) and those (hopefully few) values of x_n where $|e(n)|$ is larger than the tolerance. Since $2p$ is typically much less than N, significant data compression can be achieved. The other blocks of data can be handled similarly, with possibly different values of p.

Role of Least Squares.   To find the coefficients $a_1, \ldots, a_p$, we use Theorem 0.35. We start by putting equation (0.9), for $n = p + 1, \ldots N$, in matrix form:

$$E = Y - ZV$$

where

$$E = \begin{pmatrix} e(p+1) \\ \vdots \\ e(N) \end{pmatrix} \quad Y = \begin{pmatrix} x_{p+1} \\ \vdots \\ x_N \end{pmatrix} \quad Z = \begin{pmatrix} x_p & x_{p-1} & \cdots & x_1 \\ x_{p+1} & x_p & \cdots & x_2 \\ \vdots & & \ddots & \vdots \\ x_{N-1} & x_{N-2} & \cdots & x_{N-p} \end{pmatrix}$$

and

$$V = \begin{pmatrix} a_1 \\ \vdots \\ a_p \end{pmatrix}.$$

We want to choose $V = (a_1, \ldots, a_p)^T$ so that $\|E\|$ is as small as possible, or, in other words, so that ZV is as close as possible to Y. From Theorem 0.35, $V = (a_1, \ldots, a_p)^T$ is found by solving the following (real) matrix equation:

$$Z^T Y = Z^T Z V.$$

Written out in detail, this equation is

$$\begin{pmatrix} \langle Z_p, Y \rangle \\ \vdots \\ \langle Z_1, Y \rangle \end{pmatrix} = \begin{pmatrix} \cdots Z_p^T \cdots \\ \vdots \\ \cdots Z_1^T \cdots \end{pmatrix} \begin{pmatrix} \vdots & & \vdots \\ Z_p & \cdots & Z_1 \\ \vdots & & \vdots \end{pmatrix} \begin{pmatrix} a_1 \\ \vdots \\ a_p \end{pmatrix} \tag{0.10}$$

where we have labeled the columns of the matrix of Z by $Z_p, \ldots, Z_1$ (reverse order). The horizontal dots on either side of the Z_i^T indicate that these entries are row vectors. Likewise, the vertical dots above and below the Z_i indicate that these entries are column vectors.

Equation (0.10) is a $p \times p$ system of equations for the $a_1, \ldots, a_p$ that can be solved in terms of the Z-vectors (i.e., the original signal points, x) via Gaussian elimination.

Summary of Linear Predictive Coding

Linear predictive coding involves the following procedure.

1. Sender cuts the data into blocks

$$\{x_1 \ldots x_N\}, \{x_{N+1} \ldots x_{2N}\}, \{x_{2N+1} \ldots x_{3N}\} \ldots$$

   where each block has some near repetitive pattern. Then choose p close to the length of the repetitive pattern for the first block.

2. For $1 \le i \le p$, form the vectors

$$Z_i = \begin{pmatrix} x_i \\ \vdots \\ x_{N+i-p-1} \end{pmatrix}.$$

3. Sender solves the system of equations (0.10) for the coefficients $a_1, \ldots, a_p$ and transmits to the receiver both $a_1, \ldots, a_p$ and $x_1, \ldots, x_p$.

4. The receiver then reconstructs $x_{p+1}, \ldots, x_N$ (in this order) via the equation

$$x_n = a_1 x_{n-1} + \cdots + a_p x_{n-p} \quad (p+1 \le n \le N)$$

   for those x_n where the corresponding least squares errors, $e(n)$, are smaller than some specified tolerance. If $e(n)$ is larger than the tolerance, then the sender must transmit x_n.

Certainly, some work is required for the sender to solve the preceding equations for the $a_1, \ldots, a_p$ and for the receiver to reconstruct the x_n. You may wonder whether this work is more than the energy required to transmit all the x_i. However, keep in mind that the work required to solve for the a_i and reconstruct the x_n is done by the sender and receiver with computers that can do millions or billions of operations per second, whereas the transmission lines may only handle thousands of data bits per second. So the goal is to shift, as much as possible, the burden from the relatively slow process of transmitting the data to the much faster process of performing computations by the computers located at either the sender or the receiver.

0.8   Exercises

1. Verify that the function

$$\langle Z, W \rangle = \sum_{j=1}^n Z_j \overline{W_j}$$

   for $Z = (Z_1, \ldots, Z_n)$, $W = (W_1, \ldots, W_n) \in C^n$ defines an inner product on C^n (i.e., satisfies Definition 0.1).

2. Verify that the functions $\langle,\rangle$ defined in Examples 0.2 and 0.3, are inner products.

3. Define $\langle V, W \rangle$ for $V = (v_1, v_2)$ and $W = (w_1, w_2) \in C^2$ as

$$\langle V, W \rangle = (\overline{w_1}, \overline{w_2}) \begin{pmatrix} 1 & 2 \\ 2 & 4 \end{pmatrix} \begin{pmatrix} v_1 \\ v_2 \end{pmatrix}.$$

Show that $\langle V, V \rangle = 0$ for all vectors $V = (v_1, v_2)$ with $v_1 + 2v_2 = 0$. Does $\langle,\rangle$ define an inner product?

4. Show that the $L^2[a, b]$ inner product satisfies the following properties.

   - The L^2 inner product is conjugate symmetric (i.e., $\langle f, g \rangle = \overline{\langle g, f \rangle}$), homogeneous, and bilinear (these properties are listed in Definition 0.1).

   - Show that the L^2 inner product satisfies positivity on the space of continuous functions on $[a, b]$ by using the following outline.
     (a) We want to show that if $\int_a^b |f(t)|^2 \, dt = 0$, then $f(t) = 0$ for all $a \leq t \leq b$.
     (b) Suppose, by contradiction, that $|f(t_0)| > 0$; then use the definition of continuity to show that $|f(t)| > |f(t_0)|/2$ on an interval of the form $[t_0 - \delta, t_0 + \delta]$.
     (c) Then show

     $$\int_a^b |f(t)|^2 \, dt \geq \frac{|f(t_0)|^2}{4} [2\delta] > 0,$$

     which contradicts the assumption that $\int_a^b |f(t)|^2 \, dt = 0$.

5. Show that $\langle x, y \rangle = \sum_{n=0}^{\infty} x_n \overline{y_n}$ defines an inner product on l^2.

6. For $n > 0$, let

$$f_n(t) = \begin{cases} 1, & 0 \leq t \leq 1/n \\ 0, & \text{otherwise.} \end{cases}$$

Show that $f_n \to 0$ in $L^2[0, 1]$. Show that f_n does *not* converge to zero uniformly on $[0, 1]$.

7. For $n > 0$, let

$$f_n(t) = \begin{cases} \sqrt{n} & 0 \leq t \leq 1/n^2 \\ 0 & \text{otherwise.} \end{cases}$$

Show that $f_n \to 0$ in $L^2[0, 1]$ but that $f_n(0) \to \infty$ as $n \to \infty$.

8. Is Theorem 0.10 true on an infinite interval such as $[0, \infty)$?

9. Compute the orthogonal complement of the space in R^3 spanned by the vector $(1, -2, 1)$.

10. Let $f(t) = 1$ on $0 \le t \le 1$. Show that the orthogonal complement of f in $L^2[0,1]$ is the set of all functions whose average value is zero.

11. Show that if a differentiable function, f, is orthogonal to $\cos(t)$ on $L^2[0, \pi]$, then f' is orthogonal to $\sin(t)$ in $L^2[0, \pi]$. *Hint:* Integrate by parts.

12. By using Gram-Schmidt, find an orthonormal basis for the subspace of $L^2[0,1]$ spanned by $1, x, x^2, x^3$.

13. Find the $L^2[0,1]$ projection of the function $\cos x$ onto the space spanned by $1, x, x^2, x^3$.

14. Find the $L^2[-\pi, \pi]$ orthogonal projection of the function $f(x) = x^2$ onto the space $V_n \subset L^2[-\pi, \pi]$ spanned by

$$\left\{ 1, \frac{\sin(jx)}{\sqrt{\pi}}, \frac{\cos(jx)}{\sqrt{\pi}}; \; j = 1, \ldots, n \right\}$$

for $n = 1$. Repeat this exercise for $n = 2$ and $n = 3$. Plot these projections along with f using a computer algebra system. Repeat for $g(x) = x^3$.

15. Project the function $f(x) = x$ onto the space spanned by $\phi(x), \psi(x), \psi(2x)$, $\psi(2x - 1) \in L^2[0,1]$ where

$$\phi(x) = \begin{cases} 1, & 0 \le x < 1 \\ 0, & \text{otherwise} \end{cases} \qquad \psi(x) = \begin{cases} 1, & 0 \le x < 1/2 \\ -1, & 1/2 \le x < 1 \\ 0, & \text{otherwise.} \end{cases}$$

16. Let $D = \{(x, y) \in R^2; \; x^2 + y^2 \le 1\}$. Let

$$L^2(D) = \left\{ f : D \to C; \; \iint_D |f(x,y)|^2 \, dx \, dy < \infty \right\}.$$

Define an inner product on $L^2(D)$ by

$$\langle f, g \rangle = \iint_D f(x,y)\overline{g(x,y)} \, dx \, dy.$$

Let $\phi_n(x,y) = (x + iy)^n$, $n = 0, 1, 2, \ldots$. Show that this collection of functions is orthogonal in $L^2(D)$ and compute $\|\phi_n\|$. *Hint:* Use polar coordinates.

17. Suppose u_0 and u_1 are vectors in the inner product space V with $\langle u_0, v \rangle = \langle u_1, v \rangle$ for all $v \in V$. Show that $u_0 = u_1$. *Hint:* Let $v = u_0 - u_1$.

18. Suppose A is an $n \times n$ matrix with complex entries. Show that the following are equivalent.

   (a) The rows of A form an orthonormal basis in C^n.

(b) $AA^* = I$ (the identity matrix).

(c) $\|Ax\| = \|x\|$ for all vectors $x \in C^n$.

19. Suppose $K(x, y)$ is a continuous function that vanishes outside a bounded set in $R \times R$. Define $T : L^2(R) \to L^2(R)$ by

$$T(f)(x) = \int_{y \in R} f(y) K(x, y) \, dy.$$

Show $T^* g(x) = \int_{y \in R} \overline{K(y, x)} g(y) \, dy$. Note the parallel with the adjoint of a matrix ($A_{ij}^* = \overline{A}_{ji}$).

20. Suppose $A : V \to W$ is a linear map between two inner product spaces. Show that $\text{Ker}(A^*) = (\text{Range } A)^\perp$. Note: Ker stands for Kernel; $\text{Ker}(A^*)$ is the set of all vectors in W that are mapped to zero by A^*.

21. Prove the following theorem (Fredholm's alternative): Suppose $A : V \to W$ is a linear map between two inner product spaces. Let b be any element in W. Then either

   • $Ax = b$ has a solution for some $x \in V$, or

   • There is a vector $w \in W$ with $A^* w = 0$ and $\langle b, w \rangle_W \neq 0$.

22. Suppose V_0 is a finite dimensional subspace of an inner product space, V. Show that $V_0 = ((V_0)^\perp)^\perp$. *Hint*: The inclusion $\subset$ is easy; for the reverse inclusion, take any element $w \in ((V_0)^\perp)^\perp$ and then use Theorem 0.25 to decompose w into its components in V_0 and $V_0^\perp$. Show that its $V_0^\perp$-component is zero.

23. Show that a set of orthonormal vectors is linearly independent.

24. Verify the formulas for m and b given in Theorem 0.34.

25. Prove the uniqueness part of Theorem 0.35; *Hint*: See the proof of the uniqueness part of Theorem 0.34.

26. Obtain an alternative proof (using calculus) of Theorem 0.34 by using the following outline.

   (a) Show that the least squares problem is equivalent to finding m and b to minimize the error quantity

   $$E(m, b) = \sum_{i=1}^{N} |mx_i + b - y_i|^2.$$

   (b) From calculus, show that this minimum occurs when

   $$\frac{\partial E}{\partial m} = \frac{\partial E}{\partial b} = 0.$$

(c) Solve these two equations for m and b.

27. Obtain the best fit least squares line for these data:

x	0	1	3	4
y	0	8	8	20

28. Repeat the previous problem with the best fit least squares parabola.

29. This exercise is best done with MATLAB. The goal of this exercise is to use linear predictive coding to compress strings of numbers. Choose $X = (x_1, \ldots, x_N)$, where x_j a is periodic sequence of period p and length N. For example, $x_j = \sin(j\pi/3)$ for $1 \le j \le N = 60$ is a periodic sequence of length $p = 6$. Apply the linear predictive coding scheme to compute $a_1, \ldots, a_p$. Compute the residual $E = Y - ZV$. If done correctly, this residual should be theoretically zero (although the use of a computer will introduce a small round-off error). Now perturb X by a small randomly generated sequence [in MATLAB, add rand(1,60) to X]. Then reapply linear predictive coding and see how many terms in the residual E are small (say less than 0.1). Repeat with other sequences X on your own.

Chapter 1

Fourier Series

1.1 Introduction

In this chapter, we examine the trigonometric expansion of a function $f(x)$ defined on an interval such as $-\pi \leq x \leq \pi$. A trigonometric expansion is a sum of the form

$$a_0 + \sum_k a_k \cos(kx) + b_k \sin(kx) \tag{1.1}$$

where the sum could be finite or infinite. Why should we care about expressing a function in such a way? As the following sections show, the answer varies depending on the application we have in mind.

1.1.1 Historical Perspective

Trigonometric expansions arose in the 1700s, in connection with the study of vibrating strings and other similar physical phenomena; they became part of a controversy over what constituted a general solution to such problems, but they were not investigated in any systematic way. In 1808, Fourier wrote the first version of his celebrated memoir on the theory of heat *Théorie Analytique de la Chaleur*, which was not published until 1822. In it, he made a detailed study of trigonometric series, which he used to solve a variety of heat conduction problems.

Fourier's work was controversial at the time, partly because he *did* make unsubstantiated claims and overstate the scope of his results, but mostly because his point of view was new and strange to mathematicians of the day. For instance, in the early nineteenth century, a function was considered to be any expression involving known terms, such as powers of x, exponential functions, and trigonometric functions. The more abstract definition of a function (i.e., as a rule that assigns numbers from one set, called the domain, to another set, called the range) did not come until later. Nineteenth-century mathematicians tried to answer the following question: Can a curve in the plane, which has the property that each vertical line intersects the curve at most once, be described as the graph of a function that can be expressed using powers of x, exponentials, and trigonometric functions? In fact, they showed that for most curves, only trigonometric sums of the type given in (1.1) are needed (powers of x, exponentials, and other types of mathematical expressions are unnecessary). We shall prove this result in Theorem 1.22.

The Riemann integral and the Lebesgue integral arose in the study of Fourier series. Applications of Fourier series (and the related Fourier transform) include probability and statistics, signal processing, and quantum mechanics. Nearly two centuries after Fourier's work, the series that bears his name is still important, practically and theoretically, and still a topic of current research. For a fine historical summary and further references, see John J. Benedetto's book [1].

1.1.2   Signal Analysis

There are many practical reasons for expanding a function as a trigonometric sum. If $f(t)$ is a signal, (for example, a time-dependent electrical voltage or the sound coming from a musical instrument), then a decomposition of f into a trigonometric sum gives a description of its component frequencies. Here, we let t be the independent variable (representing time) instead of x. A sine wave, such as $\sin(kt)$, has a period of $2\pi/k$ and a frequency of k (i.e., vibrates k times in the interval $0 \le t \le 2\pi$). A signal such as

$$2\sin(t) - 50\sin(3t) + 10\sin(200t)$$

contains frequency components that vibrate at 1, 3, and 200 times per 2π-interval length. In view of the size of the coefficients, the component vibrating at a frequency of 3 dominates over the other frequency components.

A common task in signal analysis is the elimination of high-frequency noise. One approach is to express f as a trigonometric sum

$$f(t) = a_0 + \sum_k a_k \cos(kt) + b_k \sin(kt)$$

and then set the high-frequency coefficients (the a_k and b_k for large k) equal to zero.

Another common task in signal analysis is data compression. The goal here is to send a signal in a way that requires minimal data transmission. One approach is to express the signal, f, in terms of a trigonometric expansion, as previously, and then send only those coefficients, a_k and b_k, that are larger (in absolute value) than some specified tolerance. The coefficients that are small and do not contribute substantially to f can be thrown away. There is no danger that an infinite number of coefficients stay large, because we will show (see the Riemann-Lebesgue lemma, Theorem 1.21) that a_k and b_k converge to zero as $k \to \infty$.

1.1.3 Partial Differential Equations

Trigonometric sums also arise in the study of partial differential equations. Although the subject of partial differential equations is not the main focus of this book, we digress to give a simple yet important example. Consider the heat equation

$$
\begin{aligned}
u_t(x,t) &= u_{xx}(x,t) & t &> 0,\ 0 \le x \le \pi \\
u(x,0) &= f(x) & 0 &\le x \le \pi \\
u(0,t) &= A & u(\pi,t) &= B.
\end{aligned}
$$

The solution, $u(x,t)$, to this differential equation represents the temperature of a rod of length π at position x and at time t with initial temperature (at $t = 0$) given by $f(x)$ and where the temperatures at the ends of the rod, $x = 0$ and $x = \pi$, are kept at A and B, respectively. We shall compute the solution to this differential equation in the special case where $A = 0$ and $B = 0$. The expansion of f into a trigonometric series will play a crucial role in the derivation of the solution.

Separation of Variables. To solve the heat equation, we use the technique of separation of variables, which assumes that the solution is of the form

$$
u(x,t) = X(x)T(t)
$$

where $T(t)$ is a function of $t \ge 0$ and $X(x)$ is a function of x, $0 \le x \le \pi$. Inserting this expression for u into the differential equation $u_t = u_{xx}$ yields

$$
X(x)T'(t) = X''(x)T(t) \quad \text{or} \quad \frac{T'(t)}{T(t)} = \frac{X''(x)}{X(x)}.
$$

The left side depends only on t and the right side depends only on x. The only way these two functions can equal each other for all values of x and t is if both functions are constant (since x and t are independent variables). So we obtain the following two equations:

$$
\frac{T'(t)}{T(t)} = c \qquad \frac{X''(x)}{X(x)} = c
$$

where c is a constant. From the equation $T' = cT$, we obtain $T(t) = Ce^{ct}$, for some constant C. From physical considerations, the constant c must be negative [otherwise $|T(t)|$ and hence the temperature $|u(x,t)|$ would increase to infinity as $t \to \infty$]. So we write $c = -\lambda^2 < 0$ and we have $T(t) = Ce^{-\lambda^2 t}$. The differential equation for X becomes

$$X''(x) + \lambda^2 X(x) = 0, \ 0 \le x \le \pi, \qquad X(0) = 0, \ X(\pi) = 0.$$

The boundary conditions $X(0) = 0 = X(\pi)$ arise because the temperature $u(x,t) = X(x)T(t)$ is assumed to be zero at $x = 0, \ \pi$. The solution to this differential equation is

$$X(x) = a \cos(\lambda x) + b \sin(\lambda x).$$

The boundary condition $X(0) = 0$ implies that the constant a must be zero. The boundary condition $0 = X(\pi) = b \sin(\lambda \pi)$ implies that λ must be an integer, which we label k. Note that we do not want to set b equal to zero, for if both a and b were zero, the function X would be zero and hence the temperature u would be zero. This would only make sense if the initial temperature of the rod, $f(x)$, is zero.

To summarize, we have shown that the only allowable value of λ is an integer k with corresponding solutions $X_k(x) = b_k \sin(kx)$ and $T_k(t) = e^{-k^2 t}$. Each function

$$u_k(x,t) = X_k(x)T_k(t) = b_k e^{-k^2 t} \sin(kx)$$

is a solution to the heat equation and satisfies the boundary condition $u(0,t) = u(\pi,t) = 0$. The only missing requirement is the initial condition $u(x,0) = f(x)$, which we can arrange by considering the sum of the u_k:

$$u(x,t) = \sum_{k=1}^{\infty} u_k(x,t) \tag{1.2}$$

$$= \sum_{k=1}^{\infty} b_k e^{-k^2 t} \sin(kx). \tag{1.3}$$

Setting $u(x, t = 0)$ equal to $f(x)$, we obtain the equation

$$f(x) = \sum_{k=1}^{\infty} b_k \sin(kx). \tag{1.4}$$

Equation (1.4) is called a *Fourier sine expansion of f*. In the coming sections, we describe how to find such expansions (i.e., how to find the b_k). Once found, the Fourier coefficients (the b_k) can be substituted into (1.3) to give the complete solution to the heat equation.

Thus, the problem of expanding a function in terms of sines and cosines is an important one, not only from a historical perspective but also for practical problems in signal analysis and partial differential equations.

1.2 Computation of Fourier Series

1.2.1 On the Interval $-\pi \leq x \leq \pi$

In this section, we compute the Fourier coefficients, a_k and b_k, in the Fourier series

$$f(x) = a_0 + \sum_{k=1}^{\infty} a_k \cos(kx) + b_k \sin(kx).$$

We need the following result on the orthogonality of the trigonometric functions.

THEOREM 1.1   *The following integral relations hold:*

$$\frac{1}{\pi} \int_{-\pi}^{\pi} \cos(nx)\cos(kx)\,dx = \begin{cases} 1 & \text{if } n = k \geq 1 \\ 2 & \text{if } n = k = 0 \\ 0 & \text{otherwise} \end{cases} \tag{1.5}$$

$$\frac{1}{\pi} \int_{-\pi}^{\pi} \sin(nx)\sin(kx)\,dx = \begin{cases} 1 & \text{if } n = k \geq 1 \\ 0 & \text{otherwise} \end{cases} \tag{1.6}$$

$$\frac{1}{\pi} \int_{-\pi}^{\pi} \cos(nx)\sin(kx)\,dx = 0 \quad \text{for all integers } n,\, k. \tag{1.7}$$

An equivalent way of stating this theorem is that the collection

$$\left\{ \dots, \frac{\cos(2x)}{\sqrt{\pi}}, \frac{\cos(x)}{\sqrt{\pi}}, \frac{1}{\sqrt{2\pi}}, \frac{\sin(x)}{\sqrt{\pi}}, \frac{\sin(2x)}{\sqrt{\pi}}, \dots \right\} \tag{1.8}$$

is an orthonormal set of functions in $L^2([-\pi, \pi])$.

Proof   The derivations of the first two equalities use the following identities:

$$\cos((n+k)x) = \cos nx \cos kx - \sin nx \sin kx \tag{1.9}$$
$$\cos((n-k)x) = \cos nx \cos kx + \sin nx \sin kx. \tag{1.10}$$

Adding these two identities and integrating gives

$$\int_{-\pi}^{\pi} \cos nx \cos kx\,dx = \frac{1}{2} \int_{-\pi}^{\pi} \left(\cos((n+k)x) + \cos((n-k)x) \right) dx.$$

The right side can be easily integrated. If $n \neq k$, then

$$\int_{-\pi}^{\pi} \cos nx \cos kx\,dx = \frac{1}{2} \left[\frac{\sin(n+k)x}{n+k} + \frac{\sin(n-k)x}{n-k} \right]\Bigg|_{-\pi}^{\pi} = 0.$$

If $n = k \geq 1$, then

$$\int_{-\pi}^{\pi} \cos^2 nx\,dx = \int_{-\pi}^{\pi} (1/2)(1 + \cos 2nx)\,dx = \pi.$$

If $n = k = 0$, then (1.5) reduces to $(1/\pi) \int_{-\pi}^{\pi} 1 \, dx = 2$. This completes the proof of (1.5).

Equation (1.6) follows by subtracting the equations (1.9) and (1.10) and then integrating as before. Equation (1.7) follows from the fact that $\cos(nx)\sin(kx)$ is odd for $k > 0$ (see Lemma 1.7). ♦

Now we use the orthogonality relations given in (1.5) through (1.7) to compute the Fourier coefficients. We start with the equation

$$f(x) = a_0 + \sum_{k=1}^{\infty} a_k \cos(kx) + b_k \sin(kx). \tag{1.11}$$

To find a_n, we multiply both sides by $(\cos nx)/\pi$ and integrate:

$$\frac{1}{\pi} \int_{-\pi}^{\pi} f(x)\cos nx \, dx = \frac{1}{\pi} \int_{-\pi}^{\pi} \left(a_0 + \sum_{k=1}^{\infty} a_k \cos(kx) + b_k \sin(kx) \right) \cos nx \, dx.$$

From equations (1.5) through (1.7), only the cosine terms with $n = k$ contribute to the right side, and we obtain

$$\frac{1}{\pi} \int_{-\pi}^{\pi} f(x)\cos nx \, dx = a_n \quad n \geq 1.$$

Similarly, by multiplying equation (1.11) by $\sin nx$ and integrating, we obtain

$$\frac{1}{\pi} \int_{-\pi}^{\pi} f(x)\sin nx \, dx = b_n \quad n \geq 1.$$

As a special case, we compute a_0 by integrating equation (1.11) to give

$$\frac{1}{2\pi} \int_{-\pi}^{\pi} f(x) \, dx = \frac{1}{2\pi} \int_{-\pi}^{\pi} \left(a_0 + \sum_{k=1}^{\infty} a_k \cos(kx) + b_k \sin(kx) \right) dx.$$

Each sin and cos term integrates to zero and therefore

$$\frac{1}{2\pi} \int_{-\pi}^{\pi} f(x) \, dx = \frac{1}{2\pi} \int_{-\pi}^{\pi} a_0 \, dx = a_0.$$

We summarize this discussion in the following theorem.

THEOREM 1.2 *If $f(x) = a_0 + \sum_{k=1}^{\infty} a_k \cos(kx) + b_k \sin(kx)$, then*

$$a_0 = \frac{1}{2\pi} \int_{-\pi}^{\pi} f(x) \, dx \tag{1.12}$$

$$a_n = \frac{1}{\pi} \int_{-\pi}^{\pi} f(x) \cos(nx) \, dx \tag{1.13}$$

$$b_n = \frac{1}{\pi} \int_{-\pi}^{\pi} f(x) \sin(nx) \, dx. \tag{1.14}$$

The a_n and b_n are called the *Fourier coefficients* of the function f.

Remark. The crux of the proof of Theorem 1.2 is that the collection in (1.8) is orthonormal. Thus, Theorem 0.21 guarantees that the Fourier coefficients a_n and b_n are obtained by orthogonally projecting f onto the space spanned by $\cos nx$ and $\sin nx$, respectively. In fact, note that a_n and b_n are (up to a factor of $1/\pi$) the L^2 inner products of $f(x)$ with $\cos nx$ and $\sin nx$, respectively, as provided by Theorem 0.21. Thus, the preceding proof is a repeat of the proof of Theorem 0.21 for the special case of the L^2 inner product and where the orthonormal collection (the e_j) is given in (1.8).

Keep in mind that we have only shown that if f can be expressed as a trigonometric sum, then the coefficients, a_n and b_n, are given by the preceding formulas. We will show (Theorem 1.22) that most functions on the interval $[-\pi, \pi]$ can be expressed as trigonometric sums. Note that Theorem 1.2 implies that the Fourier coefficients for a given function are unique.

1.2.2 Other Intervals

Intervals of Length 2π. In Theorem 1.2, the interval of interest is $[-\pi, \pi]$. As we show in this section, Theorem 1.2 also holds for any interval of length 2π. We need the following lemma.

LEMMA 1.3 *Suppose F is any 2π-periodic function and c is any real number; then*

$$\int_{-\pi+c}^{\pi+c} F(x)\,dx = \int_{-\pi}^{\pi} F(x)\,dx. \tag{1.15}$$

Proof A simple proof of this lemma is described graphically by Figure 1. If $F \geq 0$, the left side of (1.15) represents the area under the graph of F from $x = -\pi + c$ to $x = \pi + c$ whereas the right side of (1.15) represents the area under the graph of F from $x = -\pi$ to $x = \pi$. Since F is 2π-periodic, the shaded regions in Figure 1 are the same. The process of transferring the left shaded region to the right shaded region transforms the integral on the right side of (1.15) to the left.

An analytical proof of this lemma is outlined in Exercise 21. ♦

Using this lemma with $F(x) = f(x)\cos nx$ or $f(x)\sin nx$, we see that the integration formulas in Theorem 1.2 hold for any interval of the form $[-\pi + c, \pi + c]$.

Intervals of General Length. We can also consider intervals of the form $-a \leq x \leq a$, of length $2a$. The basic building blocks are $\cos(n\pi x/a)$ and $\sin(n\pi x/a)$, which have period $2a$. Note that when $a = \pi$, these functions reduce to $\cos nx$ and $\sin nx$, which form the basis for Fourier series on the interval $[-\pi, \pi]$ considered in Theorem 1.2.

The following scaling argument can be used to transform the integral formulas for the Fourier coefficients on the interval $[-\pi, \pi]$ to the interval $[-a, a]$.

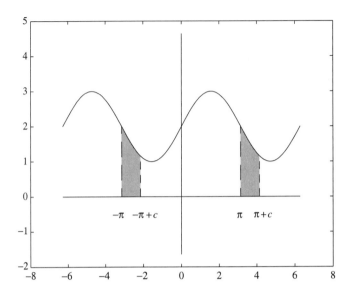

Figure 1 Region between $-\pi$ and $-\pi + c$ has same area
as between π and $\pi + c$

Suppose F is a function defined on the interval $-\pi \leq x \leq \pi$. The substitution
$x = t\pi/a$, $dx = \pi dt/a$ leads to the following change of variables formula:

$$\frac{1}{\pi} \int_{-\pi}^{\pi} F(x)\, dx = \frac{1}{a} \int_{-a}^{a} F(\frac{\pi t}{a})\, dt.$$

By using this change of variables, the following theorem can be derived from
Theorem 1.2. (See Exercise 13.)

THEOREM 1.4 *If $f(x) = a_0 + \sum_{k=1}^{\infty} a_k \cos(k\pi x/a) + b_k \sin(k\pi x/a)$ on the
interval $-a \leq x \leq a$, then*

$$a_0 = \frac{1}{2a} \int_{-a}^{a} f(t)\, dt$$

$$a_n = \frac{1}{a} \int_{-a}^{a} f(t) \cos(n\pi t/a)\, dt$$

$$b_n = \frac{1}{a} \int_{-a}^{a} f(t) \sin(n\pi t/a)\, dt.$$

EXAMPLE 1.5

Let

$$f(x) = \begin{cases} 1 & \text{if } 0 \leq x \leq 1 \\ 0 & \text{otherwise.} \end{cases}$$

We will compute the formal Fourier series for f valid on the interval $-2 \leq x \leq 2$. With $a = 2$ in Theorem 1.4, the Fourier cosine coefficients are

$$a_0 = \frac{1}{4} \int_{-2}^{2} f(t)\, dt = \frac{1}{4} \int_{0}^{1} 1\, dt = \frac{1}{4}$$

and for $n \geq 1$

$$a_n = \frac{1}{2} \int_{-2}^{2} f(t) \cos n\pi t/2\, dt = \frac{1}{2} \int_{0}^{1} \cos n\pi t/2\, dt = \frac{\sin(n\pi/2)}{n\pi}.$$

When n is even, these coefficients are zero. When $n = 2k + 1$ is odd, then $\sin(n\pi/2) = (-1)^k$. Therefore,

$$a_n = \frac{(-1)^k}{(2k+1)\pi}, \quad (n = 2k+1).$$

Similarly,

$$b_n = \frac{1}{2} \int_{-2}^{2} f(t) \sin n\pi t/2\, dt = \frac{1}{2} \int_{0}^{1} \sin n\pi t/2\, dt = \frac{-1}{n\pi}\left(\cos n\pi/2 - 1\right)$$

$$
\begin{cases}
\text{when } n = 4j,\ b_n = 0 \\[2ex]
\text{when } n = 4j + 1,\ b_n = \dfrac{1}{(4j+1)\pi} \\[2ex]
\text{when } n = 4j + 2,\ b_n = \dfrac{1}{(2j+1)\pi} \\[2ex]
\text{when } n = 4j + 3 \text{ get } \dfrac{1}{(4j+3)\pi}.
\end{cases}
$$

Thus, the Fourier series for f is

$$F(x) = a_0 + \sum_{n=1}^{\infty} a_n \cos(n\pi x/2) + b_n \sin(n\pi x/2)$$

with a_n, b_n given as above.

In later sections, we take up the question of whether or not the Fourier series $F(x)$ for f equals $f(x)$ itself. ∎

1.2.3 Cosine and Sine Expansions

Even and Odd Functions

DEFINITION 1.6 *Let $f : R \to R$ be a function; f is even if $f(-x) = f(x)$; f is odd if $f(-x) = -f(x)$.*

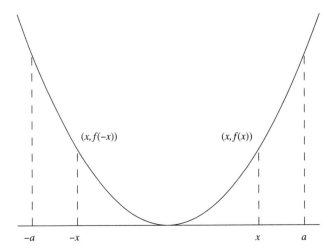

Figure 2 Even function: $f(-x) = f(x)$

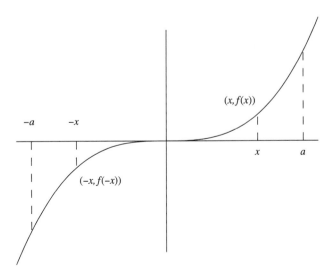

Figure 3 Odd function: $f(-x) = -f(x)$

The graph of an even function is symmetric about the y-axis, as illustrated in Figure 2. Examples include $f(x) = x^2$ (or any even power) and $f(x) = \cos x$. The graph of an odd function is symmetric about the origin, as illustrated in Figure 3. Examples include $f(x) = x^3$ (or any odd power) and $f(x) = \sin x$.

The following properties follow from the definition.

$$\text{Even} \times \text{Even} = \text{Even}$$
$$\text{Even} \times \text{Odd} = \text{Odd}$$
$$\text{Odd} \times \text{Odd} = \text{Even}.$$

For example, if f is even and g is odd, then $g(-x)f(-x) = -g(x)f(x)$ and so fg is odd.

Another important property of even and odd functions is given in the next lemma.

LEMMA 1.7

- If F is an even function, then

$$\int_{-a}^{a} F(x)\, dx = 2 \int_{0}^{a} F(x)\, dx.$$

- If F is an odd function, then

$$\int_{-a}^{a} F(x)\, dx = 0.$$

This lemma follows easily from Figures 2 and 3. If F is even, then the integral over the left half interval $[-a, 0]$ is the same as the integral over the right half interval $[0, a]$. Thus, the integral over $[-a, a]$ is twice the integral over $[0, a]$. If F is odd, then the integral over the left half interval $[-a, 0]$ cancels with the integral over the right half interval $[0, a]$. In this case, the integral over $[-a, a]$ is zero. ♦

If the Fourier series of a function only involves the cosine terms, then it must be an even function (since cosine is even). Likewise, a Fourier series that only involves sines must be odd. The converse of this fact is also true, which is the content of the next theorem.

THEOREM 1.8

- If $f(x)$ is an even function, then its Fourier series on the interval $[-a, a]$ will only involve cosines. That is, $f(x) = a_0 + \sum_{k=1}^{\infty} a_k \cos(k\pi x/a)$ with

$$a_0 = \frac{1}{a} \int_{0}^{a} f(x)\, dx$$

$$a_k = \frac{2}{a} \int_{0}^{a} f(x)\cos(k\pi x/a)\, dx.$$

- If $f(x)$ is an odd function, then its Fourier series will only involve sines. That is, $f(x) = \sum_{k=1}^{\infty} b_k \sin(k\pi x/a)$ with

$$b_k = \frac{2}{a} \int_{0}^{a} f(x)\sin(k\pi x/a)\, dx.$$

Proof   This theorem follows from Lemma 1.7 and Theorem 1.2. If f is even, then $f(x)\cos n\pi x/a$ is even and so its integral over $[-a, a]$ equals twice the integral over $[0, a]$. In addition, $f(x)\sin n\pi x/a$ is odd and so its integral over $[-a, a]$ is zero. The second part follows similarly.   ♦

Fourier Cosine and Sine Series on a Half-Interval.   Suppose f is defined on the interval $[0, a]$. By considering even or odd extensions of f, we can expand f as a cosine or sine series. To express f as a cosine series, we consider the even extension of f:

$$f_e(x) = \begin{cases} f(x) & \text{if } 0 \le x \le a \\ f(-x) & \text{if } -a \le x < 0. \end{cases}$$

The function f_e is an even function defined on $[-a, a]$. Therefore, only cosine terms appear in its Fourier expansion:

$$f_e(x) = a_0 + \sum_{k=1}^{\infty} a_k \cos k\pi x/a \quad -a \le x \le a \tag{1.16}$$

where a_k is given in Theorem 1.8. Since $f_e(x) = f(x)$ for $0 \le x \le a$, the integral formulas in Theorem 1.8 only involve $f(x)$ rather than $f_e(x)$ and so (1.16) becomes

$$f(x) = a_0 + \sum_{k=1}^{\infty} a_k \cos k\pi x/a \quad 0 \le x \le a$$

with

$$a_0 = \frac{1}{a} \int_0^a f(x)\, dx$$

$$a_k = \frac{2}{a} \int_0^a f(x)\cos(k\pi x/a)\, dx.$$

Likewise, to express f as a sine series, then we consider the odd extension of f:

$$f_o(x) = \begin{cases} f(x) & \text{if } 0 \le x \le a \\ -f(-x) & \text{if } -a \le x < 0. \end{cases}$$

The odd function, f_o, has only sine terms in its Fourier expansion. Since $f_o(x) = f(x)$ for $0 \le x \le a$, we obtain the following sine expansion for f:

$$f(x) = \sum_{k=1}^{\infty} b_k \sin k\pi x/a \quad 0 < x \le a$$

where b_k is given in Theorem 1.8:

$$b_k = \frac{2}{a} \int_0^a f(x)\sin(k\pi x/a)\, dx.$$

The examples in the next section will clarify these ideas.

1.2.4 Examples

Let f be a function and let $F(x)$ be its Fourier series on $[-\pi, \pi]$:

$$F(x) = a_0 + \sum_{n=1}^{\infty} a_n \cos nx + b_n \sin nx$$

$$= a_0 + \lim_{N \to \infty} \sum_{n=1}^{N} a_n \cos nx + b_n \sin nx$$

where a_n and b_n are the Fourier coefficients of f. We say that the Fourier series *converges* if the preceding limit exists (as $N \to \infty$). Theorems 1.2 and 1.4 only compute the Fourier series of a given function. We have not yet shown that a given Fourier series converges (or what it converges to). In Theorems 1.22 and 1.28, we show that under a mild hypothesis on the differentiability of f, the following principle holds:

Let f be a 2π-periodic function.

- If f is continuous at a point x, then its Fourier series, $F(x)$, converges and $F(x) = f(x)$.

- If f is not continuous at a point x, then $F(x)$ converges to the average of the left and right limits of f at x; that is,

$$F(x) = \frac{1}{2} \left(\lim_{t \to x^-} f(t) + \lim_{t \to x^+} f(t) \right).$$

The second statement includes the first because if f is continuous at x, then the left and right limits of f are both equal to $f(x)$ and so in this case, $F(x) = f(x)$.

Rigorous statements and proofs of Theorems 1.22 and 1.28 are given in the following sections. In this section, we present several examples to gain insight into the computation of Fourier series and the rate at which Fourier series converge.

EXAMPLE 1.9

Consider the function $f(x) = x$ on $-\pi \leq x < \pi$. This function is odd and so only the sine coefficients are nonzero. Its Fourier coefficients are

$$b_k = \frac{1}{\pi} \int_{-\pi}^{\pi} x \sin(kx)\, dx$$

$$= \frac{2(-1)^{k+1}}{k} \qquad \text{(using integration by parts)}$$

and so its Fourier series for the interval $[-\pi, \pi]$ is

$$F(x) = \sum_{k=1}^{\infty} \frac{2(-1)^{k+1}}{k} \sin(kx).$$

The function $f(x) = x$ is not 2π-periodic. Its periodic extension, $\tilde{f}$, is given in Figure 4. According to the preceding principle, $F(x)$ converges to $\tilde{f}(x)$, at points where $\tilde{f}$ is continuous. At points of discontinuity of $\tilde{f}$, $(x = \cdots -\pi, \pi, \ldots)$, $F(x)$ will converge to the average of the left and right limits of $\tilde{f}(x)$. For example, $F(\pi) = 0$ (since $\sin k\pi = 0$), which is the average of the left and right limit of $\tilde{f}$ at $x = \pi$.   ∎

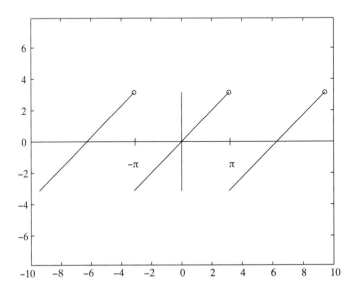

Figure 4 The 2π-periodic extension of $f(x) = x$

To see how fast the partial sums of this Fourier series converges to $\tilde{f}(x)$, we graph the partial sum

$$S_N(x) = \sum_{k=1}^{N} \frac{2(-1)^{k+1}}{k} \sin(kx)$$

for various values of N. The graph of

$$S_{10}(x) = \sum_{k=1}^{10} \frac{2(-1)^{k+1}}{k} \sin(kx)$$

is given in Figure 5 together with the graph of $\tilde{f}(x)$ (the squiggly curve is the graph of S_{10}).

First, notice that the accuracy of the approximation of $\tilde{f}(x)$ by $S_{10}(x)$ gets worse the closer x is to a point of discontinuity. For example, near $x = \pi$, the graph of $S_{10}(x)$ must travel from about $y = \pi$ to $y = -\pi$ in a very short interval, resulting in a slow rate of convergence near $x = \pi$.

Second, notice the blips in the graph of the Fourier series just before and just after the points of discontinuity of $f(x)$ (near $x = \pi$, for example). This

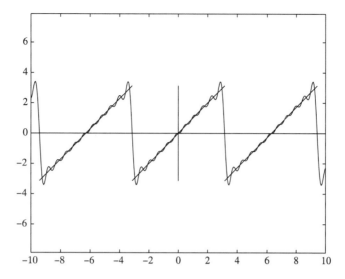

Figure 5 Gibbs phenomenon for S_{10}

effect is called the *Gibbs phenomenon*. An interesting fact about the Gibbs phenomenon is that the height of the blip is approximately the same no matter how many terms are considered in the partial sum. However, the width of the blip gets smaller as the number of terms increase. Figure 6 illustrates the Gibbs phenomenon for S_{50} (the first 50 terms of the Fourier expansion) for $\tilde{f}$. Exercise 28 explains the Gibbs effect in more detail.

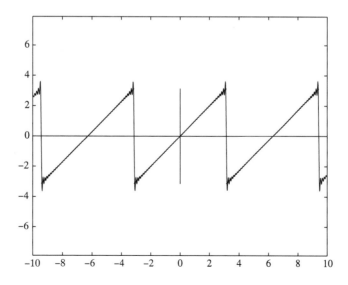

Figure 6 Gibbs phenomenon for S_{50} in Example 1.9

EXAMPLE 1.10

Consider the sawtooth wave illustrated in Figure 7. The formula for f on the interval $0 \le x \le \pi$ given by

$$f(t) = \begin{cases} x & \text{if } 0 \le x \le \pi/2 \\ \pi - x & \text{if } \pi/2 \le x \le \pi \end{cases}$$

and extends to the interval $-\pi \le x \le 0$ as an even function (see Figure 7). Since f is an even function, only the cosine terms are nonzero. Using Theorem 1.8, their coefficients are

$$a_0 = \frac{1}{\pi} \int_0^\pi f(x)\, dx = \frac{\pi}{4} \qquad \text{(no integration is needed)}$$

For $j > 0$

$$a_j = \frac{2}{\pi} \int_0^\pi f(x)\cos(jx)\, dx$$

$$= \frac{2}{\pi} \int_0^{\frac{\pi}{2}} x \cos(jx)\, dx + \frac{2}{\pi} \int_{\frac{\pi}{2}}^\pi (\pi - x)\cos(jx)\, dx.$$

After performing the necessary integrals, we have

$$a_j = \frac{4\cos(j\pi/2) - 2\cos(j\pi) - 2}{\pi j^2} \qquad \text{for } j > 0.$$

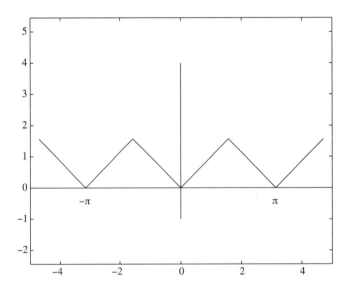

Figure 7 Sawtooth wave

Only the a_{4k+2} are nonzero. These coefficients simplify to

$$a_{4k+2} = \frac{-2}{\pi(2k+1)^2}.$$

So the Fourier series for the sawtooth wave is

$$F(x) = \frac{\pi}{4} - \frac{2}{\pi}\sum_{k=0}^{\infty}\frac{1}{(2k+1)^2}\cos((4k+2)x). \qquad \blacksquare$$

The sawtooth wave is already periodic and it is continuous. Thus its Fourier series $F(x)$ equals the sawtooth wave, $f(x)$, for every x by the principle stated at the beginning of this section. In addition, the rate of convergence is much faster than for the Fourier series in Example 1.9. To illustrate the rate of convergence, we plot the sum of the first two terms of its Fourier series

$$S_2(x) = \frac{\pi}{4} - \frac{2\cos(2x)}{\pi}$$

in Figure 8.

The sum of just two terms of this Fourier series gives a more accurate approximation of the sawtooth wave than 10 or 50 or even 1000 terms of the Fourier series in the previous (discontinuous) example. Indeed, the graph of the first 10 terms of this Fourier series (given in Figure 9) is almost indistinguishable from the original function.

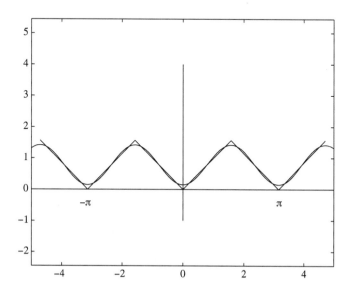

Figure 8 Sum of first two terms of the Fourier series of the sawtooth

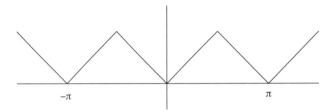

Figure 9 Ten terms of the Fourier series of the sawtooth

EXAMPLE 1.11

Let $f(x) = \sin(3x) + \cos(4x)$. Since f is already expanded in terms of sines and cosines, no work is needed to compute the Fourier series of f [i.e., the Fourier series of f is just $\sin(3x) + \cos(4x)$]. This example illustrates an important point. The Fourier coefficients are unique (Theorem 1.2 specifies exactly what the a_k and b_k must be). Thus by inspection, $b_3 = 1$, $a_4 = 1$, and all other a_k and b_k are zero. By uniqueness, these are the same values as would have been obtained by computing the integrals in Theorem 1.2 for the a_k and b_k (but with much less work). ∎

EXAMPLE 1.12

Let $f(x) = \sin^2(x)$. In this example, f is not written as a linear combination of sines and cosines, so there is some work to do. However, instead of computing the integrals in Theorem 1.2 for the a_k and b_k, we use a trigonometric identity

$$\sin^2(x) = \frac{1}{2}(1 - \cos(2x)).$$

The right side is the desired Fourier series for f since it is a linear combination of $\cos(kx)$ (here, $a_0 = 1/2$, $a_2 = -1/2$ and all other a_k and b_k are zero). ∎

EXAMPLE 1.13

To find the Fourier sine series for the function $f(x) = x^2 + 1$ valid on the interval $0 \le x \le 1$, we first extend f as an odd function:

$$f_o(x) = \begin{cases} f(x) = x^2 + 1 & \text{for } 0 \le x \le 1 \\ -f(-x) = -x^2 - 1 & \text{for } -1 \le x \le 0. \end{cases}$$

Then we use Theorem 1.8 to compute the Fourier coefficients for f_o.

$$b_n = 2\int_0^1 f(x)\sin(n\pi x)\,dx = 2\int_0^1 (x^2+1)\sin(n\pi x)\,dx.$$

Note that the formula of the odd extension of f to the interval $-1 \le x \le 0$ is not needed for the computation of b_n. Integration by parts (twice) gives

$$b_n = -2\frac{2n^2\pi^2(-1)^n - 2(-1)^n + 2 - n^2\pi^2}{\pi^3 n^3}.$$

When $n = 2k$ is even, this simplifies to

$$b_{2k} = -\frac{1}{k\pi}$$

and when $n = 2k - 1$ is odd

$$b_{2k-1} = 2\frac{12k^2\pi^2 - 12\pi^2 k + 3\pi^2 - 4}{\pi^3(2k-1)^3}.$$

Thus the Fourier sine series for $x^2 + 1$ on the interval $[0, 1]$ is

$$F(x) = \sum_{k=1}^{\infty} -\left(\frac{1}{k\pi}\right)\sin 2k\pi x + 2\left(\frac{12k^2\pi^2 - 12\pi^2 k + 3\pi^2 - 4}{\pi^3(2k-1)^3}\right)\sin(2k-1)\pi x.$$

$$(1.17)$$

Now f_o is defined on the interval $[-1, 1]$. Its periodic extension, $\tilde{f}_o$, is graphed on the interval $[-2, 2]$ in in Figure 10. Its Fourier series, $F(x)$, will converge to $\tilde{f}_o(x)$ at each point of continuity of $\tilde{f}_o$. At each integer, $\tilde{f}_o$ is discontinuous. By the principle stated at the beginning of this section, $F(x)$ will converge to zero at each integer value (the average of the left and right limits of $\tilde{f}_o$). This agrees with the value of F computed by using (1.17) [since $\sin 2k\pi$ and $\sin(2k-1)\pi$ are zero for each integer k]. A graph of $F(x)$ is given in Figure 11. Note that since $\tilde{f}_o(x) = f(x)$ for $0 < x < 1$, the Fourier sine series $F(x)$ agrees with $f(x) = x^2 + 1$ on the interval $0 < x < 1$. A partial sum of the first 30 terms of $F(x)$ is given in Figure 12. ∎

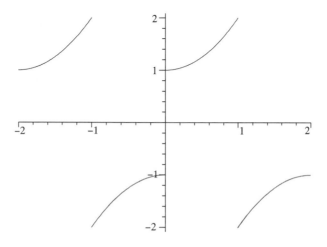

Figure 10 Periodic odd extension of $f(x) = x^2 + 1$

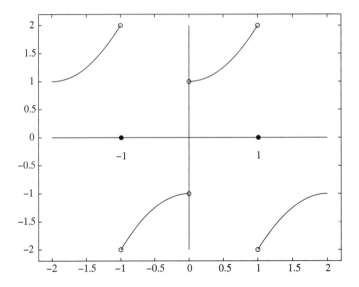

Figure 11 Graph of F, the Fourier sine series of $f(x) = x^2 + 1$

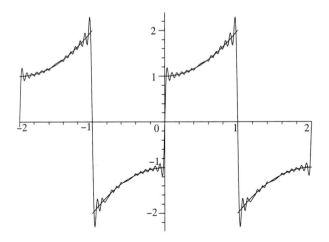

Figure 12 Graph of sum of first 30 terms of $F(x)$

EXAMPLE 1.14

Solve the heat equation

$$
\begin{aligned}
u_t(x,t) &= u_{xx}(x,t) & & t > 0, 0 \le x \le \pi \\
u(x,0) &= f(x) & & 0 \le x \le \pi \\
u(0,t) &= 0 & & u(\pi,t) = 0.
\end{aligned}
$$

where $f(x)$ is the sawtooth wave in Example 1.10; that is,

$$f(x) = \begin{cases} x & \text{if } 0 \leq x \leq \pi/2 \\ \pi - x & \text{if } \pi/2 \leq x \leq \pi. \end{cases}$$

From the discussion in Section 1.1.3, the solution to this problem is

$$u(x, t) = \sum_{k=1}^{\infty} b_k e^{-k^2 t} \sin(kx). \tag{1.18}$$

Setting $t = 0$ in (1.18) and using $u(x, 0) = f(x)$, we obtain

$$f(x) = u(x, 0) = \sum_{k=1}^{\infty} b_k \sin(kx).$$

Therefore, the b_k must be chosen as the Fourier sine coefficients of $f(x)$, which by Theorem 1.8 are

$$b_k = \frac{2}{\pi} \int_0^{\pi} f(t) \sin(kt) \, dt.$$

Inserting the formula for f and computing the integral, we can show $b_k = 0$ when k is even. When $k = 2j + 1$ is odd, then

$$b_{2j+1} = \frac{4(-1)^j}{\pi(2j+1)^2}.$$

Substituting b_k into (1.18), the final solution is

$$u(x, t) = \sum_{j=0}^{\infty} \frac{4(-1)^j}{\pi(2j+1)^2} \sin((2j+1)x) e^{-(2j+1)^2 t}. \qquad \blacksquare$$

1.2.5 The Complex Form of Fourier Series

Often, it is more convenient to express Fourier series in complex form using the complex exponentials, e^{inx}, $n \in Z$, due to the simple computational properties of these functions. The complex exponential has the following definition.

DEFINITION 1.15 *For any real number t, the complex exponential is*

$$e^{it} = \cos(t) + i \sin(t)$$

where $i = \sqrt{-1}$.

This definition is motivated by substituting $x = it$ into the usual Taylor series for e^x:

$$e^x = 1 + x + \frac{x^2}{2!} + \frac{x^3}{3!} + \frac{x^4}{4!} + \cdots$$

with $x = it$: $\quad e^{it} = 1 + (it) + \frac{(it)^2}{2!} + \frac{(it)^3}{3!} + \frac{(it)^4}{4!} + \cdots.$

Collecting the real and imaginary parts, we obtain

$$e^{it} = \left(1 - \frac{t^2}{2!} + \frac{t^4}{4!} + \dots\right) + i\left(t - \frac{t^3}{3!} + \frac{t^5}{5!} - \dots\right)$$

$$= \cos(t) + i\sin(t) \qquad \text{using the Taylor expansions of sin and cos.}$$

The next lemma shows that the familiar properties of the real exponential also hold for the complex exponential. These properties follow from the definition together with basic trigonometric identities and will be left to the exercises (see Exercise 14).

LEMMA 1.16   *For all t, $s \in R$*

- $e^{i(t+2\pi)} = e^{it}$

- $|e^{it}| = 1$

- $\overline{e^{it}} = e^{-it}$

- $e^{it}e^{is} = e^{i(t+s)}$

- $e^{it}/e^{is} = e^{i(t-s)}$

- $\frac{d}{dt}\left\{e^{it}\right\} = ie^{it}$

The next theorem shows that the complex exponentials are orthonormal in $L^2([\pi, \pi])$.

THEOREM 1.17   *The set of functions*

$$\left\{\frac{e^{int}}{\sqrt{2\pi}}, \ n = \dots, -2, -1, 0, 1, 2, \dots\right\}$$

is orthonormal in $L^2([-\pi, \pi])$.

Proof   We must show

$$\langle e^{int}, e^{imt}\rangle_{L^2} = \frac{1}{2\pi}\int_{-\pi}^{\pi} e^{int}\,\overline{e^{imt}}\,dt = \begin{cases}1 & \text{if } n = m \\ 0 & \text{if } n \neq m.\end{cases}$$

Using the third, fourth, and sixth properties in Lemma 1.16, we have

$$\int_{-\pi}^{\pi} e^{int}\overline{e^{imt}}\,dt = \int_{-\pi}^{\pi} e^{int}e^{-imt}\,dt$$

$$= \int_{-\pi}^{\pi} e^{i(n-m)t}\,dt$$

$$= \left.\frac{e^{i(n-m)t}}{i(n-m)}\right|_{-\pi}^{\pi} \qquad \text{if } n \neq m$$

$$= 0.$$

If $n = m$, then $e^{int}e^{-int} = 1$ and so $\langle e^{int}, e^{int} \rangle = 2\pi$. This completes the proof.

♦

Combining this theorem with Theorem 0.21, we obtain the following complex version of Fourier series.

THEOREM 1.18   *If $f(t) = \sum_{n=-\infty}^{\infty} \alpha_n e^{int}$ on the interval $-\pi \leq t \leq \pi$, then*

$$\alpha_n = \frac{1}{2\pi} \int_{-\pi}^{\pi} f(t)e^{-int} \, dt.$$

♦

EXAMPLE 1.19
_____

Consider the function

$$f(t) = \begin{cases} 1 & \text{if } 0 \leq t < \pi \\ -1 & \text{if } -\pi \leq t < 0. \end{cases}$$

The nth complex Fourier coefficient is

$$\alpha_n = \frac{1}{2\pi} \int_{-\pi}^{\pi} f(t)e^{-int} \, dt$$

$$= \frac{1}{2\pi} \int_{0}^{\pi} e^{-int} \, dt - \frac{1}{2\pi} \int_{-\pi}^{0} e^{-int} \, dt$$

$$= -\frac{i(1 - \cos(n\pi))}{n\pi}$$

$$= \begin{cases} \dfrac{-2i}{n\pi} & \text{if } n \text{ is odd} \\ 0 & \text{if } n \text{ is even.} \end{cases}$$

So the complex Fourier series of f is

$$\sum_{n=-\infty}^{\infty} \alpha_n e^{int} = \sum_{k=-\infty}^{\infty} \frac{-2i}{(2k+1)\pi} e^{i(2k+1)t}.$$

The complex form of Fourier series can be formulated on other intervals as well.

■

THEOREM 1.20   *The set of functions*

$$\left\{ \frac{1}{\sqrt{2a}} e^{in\pi t/a}, \ n = \cdots - 2, -1, 0, 1, 2, \ldots \right\}$$

is an orthonormal basis for $L^2[-a, a]$. If $f(t) = \sum_{n=-\infty}^{\infty} \alpha_n e^{in\pi t/a}$, then

$$\alpha_n = \frac{1}{2a} \int_{-a}^{a} f(t)e^{-in\pi t/a} \, dt.$$

Relation between the Real and Complex Fourier Series. If f is a real-valued function, the real form of its Fourier series can be derived from its complex form and vice versa. For simplicity, we discuss this derivation on the interval $-\pi \le t \le \pi$, but this discussion also holds for other intervals as well. We first decompose the complex form of the Fourier series of f into positive and negative terms:

$$f(t) = \sum_{n=-\infty}^{-1} \alpha_n e^{int} + \alpha_0 + \sum_{n=1}^{\infty} \alpha_n e^{int} \tag{1.19}$$

where

$$\alpha_n = \frac{1}{2\pi} \int_{-\pi}^{\pi} f(t) e^{-int}\, dt.$$

If f is real valued, then $\alpha_{-n} = \overline{\alpha_n}$ because

$$\alpha_{-n} = \frac{1}{2\pi} \int_{-\pi}^{\pi} f(t) e^{int}\, dt = \frac{1}{2\pi} \overline{\int_{-\pi}^{\pi} f(t) e^{-int}\, dt} = \overline{\alpha_n}.$$

Therefore, (1.19) becomes

$$f(t) = \alpha_0 + \left(\sum_{n=1}^{\infty} \alpha_n e^{int} \right) + \overline{\left(\sum_{n=1}^{\infty} \alpha_n e^{int} \right)}.$$

Since $z + \bar{z} = 2\,\mathrm{Re}(z)$ for any complex number z, this equation can be written as

$$f(t) = \alpha_0 + 2\,\mathrm{Re}\left(\sum_{n=1}^{\infty} \alpha_n e^{int} \right). \tag{1.20}$$

Now note the following relations between α_n and the real Fourier coefficients, a_n and b_n, given in Theorem 1.2:

$$\alpha_0 = \frac{1}{2\pi} \int_{-\pi}^{\pi} f(t)\, dt = a_0$$

$$\alpha_n = \frac{1}{2\pi} \int_{-\pi}^{\pi} f(t) e^{-int}\, dt \quad \text{for } n \ge 1$$

$$= \frac{1}{2\pi} \int_{-\pi}^{\pi} f(t)(\cos nt - i \sin nt)\, dt$$

$$= \frac{1}{2}(a_n - ib_n).$$

Using equation (1.20), we obtain

$$f(t) = \alpha_0 + 2\,\mathrm{Re}\left(\sum_{n=1}^{\infty} \alpha_n e^{int}\right)$$

$$= a_0 + \mathrm{Re}\left(\sum_{n=1}^{\infty}(a_n - ib_n)(\cos nt + i\sin nt)\right)$$

$$= a_0 + \sum_{n=1}^{\infty} a_n \cos nt + b_n \sin nt,$$

which is the real form of the Fourier series of f. These equations can also be stated in reverse order so that the complex form of its Fourier series can be derived from its real Fourier series.

1.3   Convergence Theorems for Fourier Series

In this section, we prove the convergence of Fourier series under mild assumptions on the original function f. The mathematics underlying convergence is somewhat involved. We start with the Riemann-Lebesgue lemma, which is important in its own right.

1.3.1   The Riemann-Lebesgue Lemma

In the examples in Section 1.2.4, the Fourier coefficients a_k and b_k converge to zero as k gets large. This is not a coincidence, as the following theorem shows.

THEOREM 1.21 Riemann-Lebesgue Lemma. *Suppose f is a piecewise continuous function on the interval $a \le x \le b$. Then*

$$\lim_{k \to \infty} \int_a^b f(x)\cos(kx)\,dx = \lim_{k \to \infty} \int_a^b f(x)\sin(kx)\,dx = 0.$$

By definition, a piecewise continuous function has only a finite number of discontinuities. This hypothesis can be weakened considerably.

One important consequence of Theorem 1.21 is that only a finite number of Fourier coefficients are larger (in absolute value) than any given positive number. This fact is the basic idea underlying the process of data compression. One way to compress a signal is first to express it as a Fourier series; then discard all the small Fourier coefficients and retain (or transmit) only the finite number of Fourier coefficients that are larger than some given threshold. (See Exercise 31 for an illustration of this process.) We encounter data compression again in future sections on the discrete Fourier transform and wavelets.

Proof The intuitive idea behind the proof is that as k gets large, $\sin(kx)$ and $\cos(kx)$ oscillate much more rapidly than does f (see Figure 13). If k is large, $f(x)$ is nearly constant on two adjacent periods of $\sin kx$ or $\cos kx$. The graph of the product of $f(x)$ with $\sin(kx)$ is given in Figure 14. The integral over each period is almost zero, since the areas above and below the x-axis almost cancel.

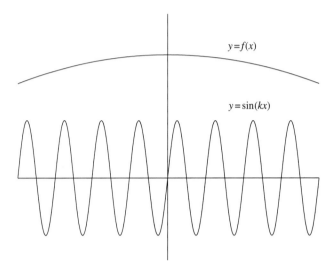

Figure 13 Plot of both $y = f(x)$ and $y = \sin kx$

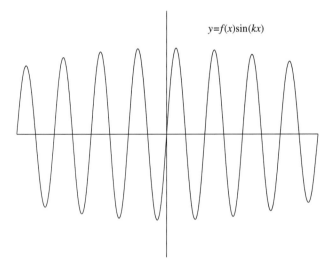

Figure 14 Plot of $y = f(x)\sin kx$

We give the following analytical proof of the theorem in the case of a differentiable function f. Consider the term

$$\int_a^b f(x)\cos(kx)\, dx.$$

We integrate by parts with $u = f$ and $dv = \cos kx$ to obtain

$$\int_a^b f(x)\cos(kx)\, dx = \frac{\sin(kx)}{k} f(x)\Big|_a^b - \int_a^b \frac{\sin(kx)}{k} f'(x)\, dx$$

$$= \frac{\sin(kb)f(b) - \sin(ka)f(a)}{k} - \int_a^b \frac{\sin(kx)}{k} f'(x)\, dx.$$

As k gets large in the denominator, the expression on the right converges to zero, and this completes the proof. ♦

1.3.2 Convergence at a Point of Continuity

The sum appearing in a Fourier expansion generally contains an infinite number of terms (the Fourier expansions of Examples 1.11 and 1.12 are an exception). An infinite sum is, by definition, a limit of partial sums; that is,

$$\sum_{k=1}^{\infty} a_k \cos(kx) + b_k \sin(kx) = \lim_{N\to\infty} \sum_{k=1}^{N} a_k \cos(kx) + b_k \sin(kx)$$

provided that the limit exists. Therefore, we say that the Fourier series of f converges to f at the point x if

$$f(x) = a_0 + \lim_{N\to\infty} \sum_{k=1}^{N} a_k \cos(kx) + b_k \sin(kx).$$

With this in mind, we state and prove our first theorem on the convergence of Fourier series.

THEOREM 1.22 *Suppose f is a continuous and 2π-periodic function. Then for each point x, where the derivative of f is defined, the Fourier series of f at x converges to $f(x)$.*

Proof For a positive integer N, let

$$S_N(x) = a_0 + \sum_{k=1}^{N} a_k \cos(kx) + b_k \sin(kx)$$

where a_k and b_k are the Fourier coefficients of the given function f. Our ultimate goal is to show $S_N(x) \to f(x)$ as $N \to \infty$. Before this can be done, we need to rewrite S_N into a different form. This process requires several steps.

Step 1. Substituting the Fourier Coefficients.   After substituting the formulas for the a_k and b_k, (1.12) through (1.14), of Theorem 1.2, we obtain

$$S_N(x) = \frac{1}{2\pi} \int_{-\pi}^{\pi} f(t)\, dt$$

$$+ \frac{1}{\pi} \sum_{k=1}^{N} \left(\int_{-\pi}^{\pi} f(t)\cos(kt)\cos(kx)\, dt + \int_{-\pi}^{\pi} f(t)\sin(kt)\sin(kx)\, dt \right)$$

$$= \frac{1}{\pi} \int_{-\pi}^{\pi} f(t) \left(\frac{1}{2} + \sum_{k=1}^{N} \cos(kt)\cos(kx) + \sin(kt)\sin(kx) \right) dt.$$

Using the addition formula for the cosine function, $\cos(A - B) = \cos(A)\cos(B) + \sin(A)\sin(B)$, we obtain

$$S_N(x) = \frac{1}{\pi} \int_{-\pi}^{\pi} f(t) \left(\frac{1}{2} + \sum_{k=1}^{N} \cos(k(t - x)) \right) dt. \tag{1.21}$$

To evaluate the sum on the right side, we need the following lemma.

Step 2. Evaluating the Fourier Kernel.

LEMMA 1.23   *For any number u, $-\pi \leq u \leq \pi$,*

$$\frac{1}{2} + \cos(u) + \cos(2u) + \cdots + \cos(Nu) = \begin{cases} \dfrac{\sin((N + 1/2)u)}{2\sin(u/2)} & u \neq 0 \\ N + 1/2 & u = 0. \end{cases}$$

Proof of Lemma 1.23   Recall that the complex exponential is defined as $e^{iu} = \cos(u) + i\sin(u)$. Note that

$$(e^{iu})^n = e^{inu} = \cos(nu) + i\sin(nu).$$

So $\cos nu = \operatorname{Re}\{(e^{iu})^n\}$. We have

$$\frac{1}{2} + \cos(u) + \cos(2u) + \cdots + \cos(Nu)$$

$$= \frac{-1}{2} + (1 + \cos(u) + \cos(2u) + \cdots + \cos(Nu))$$

and so

$$\frac{1}{2} + \sum_{k=1}^{N} \cos ku = \frac{-1}{2} + \operatorname{Re}\left\{ \sum_{k=0}^{N} (e^{iu})^k \right\}. \tag{1.22}$$

The sum on the right is a geometric series, $\sum_{k=0}^{N} z^k$, where $z = e^{iu}$.

For any number z, we have

$$\sum_{k=0}^{N} z^k = \frac{1 - z^{N+1}}{1 - z}. \qquad (1.23)$$

This formula is established as follows: Let

$$s_N = \sum_{k=0}^{N} z^k.$$

Then

$$(1 - z)s_N = (1 - z)(1 + z + z^2 + \cdots + z^N)$$
$$= (1 + z + \cdots + z^N) - (z + z^2 + \cdots + z^{N+1})$$
$$= 1 - z^{N+1}.$$

Dividing both sides by $(1 - z)$ yields (1.23).

Applying (1.23) with $z = e^{iu}$ to (1.22), we obtain

$$\frac{1}{2} + \cos(u) + \cos(2u) + \cdots + \cos(Nu) = \frac{-1}{2} + \mathrm{Re}\left\{\frac{1 - e^{i(N+1)u}}{1 - e^{iu}}\right\}. \qquad (1.24)$$

To compute the expression on the right, we multiply the numerator and denominator by $e^{-iu/2}$:

$$\mathrm{Re}\left\{\frac{1 - e^{i(N+1)u}}{1 - e^{iu}}\right\} = \mathrm{Re}\left\{\frac{e^{-iu/2} - e^{i(N+1/2)u}}{e^{-iu/2} - e^{iu/2}}\right\}.$$

The denominator on the right is $-2i\sin(u/2)$, so

$$\mathrm{Re}\left\{\frac{1 - e^{i(N+1)u}}{1 - e^{iu}}\right\} = \frac{\sin(u/2) + \sin((N + 1/2)u)}{2\sin(u/2)}.$$

Inserting this equation into the right side of (1.24) gives

$$\frac{1}{2} + \cos(u) + \cos(2u) + \cdots + \cos(Nu) = \frac{-1}{2} + \frac{\sin(u/2) + \sin((N + 1/2)u)}{2\sin(u/2)}$$
$$= \frac{\sin((N + 1/2)u)}{2\sin(u/2)}.$$

This completes the proof of Lemma 1.23.

Step 3. Evaluation of the Partial Sum of Fourier Series. Using Lemma 1.23 with $u = t - x$, (1.21) becomes

$$S_N(x) = \frac{1}{\pi} \int_{-\pi}^{\pi} f(t) \left(\frac{1}{2} + \sum_{k=1}^{N} \cos(k(t - x)) \right) dt$$

$$= \frac{1}{2\pi} \int_{-\pi}^{\pi} f(t) \left(\frac{\sin((N + 1/2)(t - x))}{\sin((t - x)/2)} \right) dt$$

$$= \int_{-\pi}^{\pi} f(t) P_N(t - x) \, dt$$

where we have let

$$P_N(u) = \frac{1}{2\pi} \frac{\sin((N + 1/2)u)}{\sin(u/2)}. \tag{1.25}$$

Now use the change of variables $u = t - x$ in the preceding integral to obtain

$$S_N(x) = \int_{-\pi - x}^{\pi - x} f(u + x) P_N(u) \, du.$$

Since both f and P_N are periodic with period 2π, the limits of integration can be shifted by x without changing the value of the integral (see Lemma 1.3). Therefore,

$$S_N(x) = \int_{-\pi}^{\pi} f(u + x) P_N(u) \, du. \tag{1.26}$$

Next, we need the following lemma.

Step 4. Integrating the Fourier Kernel.

LEMMA 1.24

$$\int_{-\pi}^{\pi} P_N(u) \, du = 1$$

Proof of Lemma 1.24   We use Lemma 1.23 to write

$$P_N(u) = \frac{1}{\pi} \frac{\sin((N + 1/2)u)}{2 \sin(u/2)}$$

$$= \frac{1}{\pi} \left(\frac{1}{2} + \cos(u) + \cos(2u) + \cdots + \cos(Nu) \right).$$

Integrating this equation gives

$$\int_{-\pi}^{\pi} P_N(u) \, du = \int_{-\pi}^{\pi} \frac{1}{2\pi} \, du + \frac{1}{\pi} \int_{-\pi}^{\pi} \cos(u) + \cos(2u) + \cdots + \cos(Nu) \, du.$$

The second integral on the right is zero (since the antiderivatives involve sin which vanishes at multiples of π). The first integral on the right is one and the proof of the lemma is complete.

Step 5. The End of the Proof of Theorem 1.22. As indicated at the beginning of the proof, we must show that $S_N(x) \to f(x)$ as $N \to \infty$. Inserting the expression for $S_N(x)$ given in (1.26), we must show

$$\int_{-\pi}^{\pi} f(u+x)P_N(u)\,du \to f(x). \tag{1.27}$$

In view of Lemma 1.24, $f(x) = \int_{-\pi}^{\pi} f(x)P_N(u)\,du$ and so we are reduced to showing

$$\int_{-\pi}^{\pi} (f(u+x) - f(x))P_N(u)\,du \to 0 \quad \text{as} \quad N \to \infty.$$

Using (1.25), the preceding limit becomes

$$\frac{1}{2\pi} \int_{-\pi}^{\pi} \left(\frac{f(u+x) - f(x)}{\sin(u/2)} \right) \sin((N+1/2)u)\,du \to 0. \tag{1.28}$$

We can use the Riemann-Lebesgue lemma (see Theorem 1.21) to establish (1.28) provided that we show that the function

$$g(u) = \frac{f(u+x) - f(x)}{\sin(u/2)}$$

is continuous (as required by the Riemann-Lebesgue lemma). Here, x is a fixed point and u is the variable. The only possible value of $u \in [-\pi, \pi]$ where $g(u)$ could be discontinuous is $u = 0$. However, since $f'(x) = \lim_{u \to 0} \frac{f(u+x)-f(x)}{u}$ exists by hypothesis, we have

$$\lim_{u \to 0} g(u) = \lim_{u \to 0} \frac{f(u+x) - f(x)}{\sin(u/2)}$$

$$= \lim_{u \to 0} \frac{f(u+x) - f(x)}{u} \frac{u/2}{\sin(u/2)} \cdot 2$$

$$= f'(x) \cdot 1 \cdot 2 \quad (\text{because } \lim_{t \to 0} \frac{t}{\sin t} = 1)$$

$$= 2f'(x).$$

Thus $g(u)$ extends across $u = 0$ as a continuous function by defining $g(0) = 2f'(x)$. We conclude that (1.28) holds by the Riemann-Lebesgue lemma. This finishes the proof of Theorem 1.22. ♦

1.3.3   Convergence at a Point of Discontinuity

Now we discuss some variations of Theorem 1.22. Note that the hypothesis of this theorem requires the function f to be continuous and periodic. However, there are many functions of interest that are neither continuous or periodic. For example, the function in Example 1.9, $f(x) = x$, is not periodic. Moreover, the periodic extension of f (graphed in Figure 15) is not continuous.

Before we state the theorem on convergence near a discontinuity, we need the following definition.

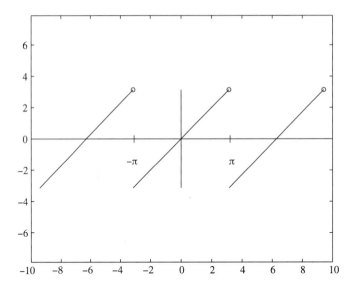

Figure 15 2π-periodic extension of $f(x) = x$

DEFINITION 1.25

- *The left and right limits of f at a point x are defined as follows.*

$$\text{Left Limit}: \quad f(x - 0) = \lim_{h \to 0^+} f(x - h)$$

$$\text{Right Limit}: \quad f(x + 0) = \lim_{h \to 0^+} f(x + h)$$

- *The function f is said to be* left differentiable *at x if the following limit exists:*

$$f'(x - 0) = \lim_{h \to 0^-} \frac{f(x + h) - f(x)}{h}.$$

- *The function f is said to be* right differentiable *at x if the following limit exists:*

$$f'(x + 0) = \lim_{h \to 0^+} \frac{f(x + h) - f(x)}{h}.$$

Intuitively, $f'(x - 0)$ represents the slope of the tangent line to f at x considering only the part of the graph of $y = f(t)$ that lies to the left of $t = x$. The value $f'(x + 0)$ is the slope of the tangent line of $y = f(t)$ at x considering only the graph of the function that lies to the right of $t = x$ (see Figure 16).

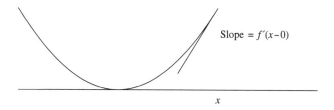

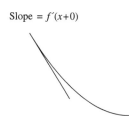

Figure 16 Left and right derivatives

EXAMPLE 1.26

Let $f(x)$ be the periodic extension of $y = x$, $-\pi \leq x < \pi$ (see Figure 15). Then $f(x)$ is discontinuous at $x = \dots, -\pi, \pi, \dots$. The left and right limits of f at $x = \pi$ are

$$f(\pi - 0) = \pi \qquad f(\pi + 0) = -\pi.$$

The left and right derivatives at $x = \pi$ are

$$f'(\pi - 0) = 1 \quad \text{and} \quad f'(\pi + 0) = 1. \qquad \blacksquare$$

EXAMPLE 1.27

Let

$$f(x) = \begin{cases} x & \text{if } 0 \leq x \leq \pi/2 \\ \pi - x & \text{if } \pi/2 \leq x \leq \pi. \end{cases}$$

The graph of f is the sawtooth wave illustrated in Figure 7. This function is continuous but not differentiable at $x = \pi/2$. The left and right derivatives at $x = \pi/2$ are

$$f'(\pi/2 - 0) = 1 \quad \text{and} \quad f'(\pi/2 + 0) = -1. \qquad \blacksquare$$

Now we are ready to state the convergence theorem for Fourier series at a point where f is not necessarily continuous.

THEOREM 1.28 *Suppose $f(x)$ is periodic and piecewise continuous. Suppose x is a point where f is left and right differentiable (but not necessarily continuous). Then the Fourier series of f at x converges to*

$$\frac{f(x+0) + f(x-0)}{2}.$$

This theorem states that at a point of discontinuity of f, the Fourier series of f converges to the average of the left and right limits of f. At a point of continuity, the left and right limits are the same, and so in this case Theorem 1.28 reduces to Theorem 1.22.

EXAMPLE 1.29

Let $f(x)$ be the periodic extension of $y = x$, $-\pi \leq x < \pi$. As mentioned in Example 1.26, f is not continuous but left and right differentiable at $x = \pi$. Theorem 1.28 states that its Fourier series, $F(x)$, converges to the average of the left and right limits of f at $x = \pi$. Since $f(\pi - 0) = \pi$ and $f(\pi + 0) = -\pi$, Theorem 1.28 implies $F(\pi) = 0$. This value agrees with the formula for the Fourier series computed in Example 1.9:

$$F(x) = \sum_{k=1}^{\infty} \frac{2(-1)^{k+1}}{k} \sin(kx)$$

whose value is zero at $x = \pi$. The graph of the F is given in Figure 17. Note that the value of F at $x = \pm\pi$ and $x = \pm 3\pi$ (indicated by the solid dots) is equal to the average of the left and right limits at $x = \pm\pi$ and $x = \pm 3\pi$.   ∎

Proof of Theorem 1.28   The proof of this theorem is very similar to the proof of Theorem 1.22. We summarize the modifications.

Steps 1 through 3 go through without change. Step 4 needs to be modified as follows

Step 4′.

$$\int_0^\pi P_N(u)\, du = \int_{-\pi}^0 P_N(u)\, du = \frac{1}{2}$$

where

$$P_N(u) = \frac{1}{2\pi} \frac{\sin(N + 1/2)u}{\sin u/2}.$$

In fact, these equalities follow from Lemma 1.24, and by using the fact that $P_N(u)$ is even (so the left- and right-half integrals are equal and sum to 1).

Step 5 is now replaced by the following.

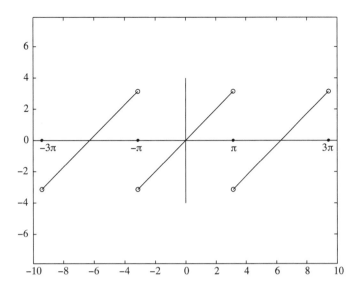

Figure 17 Graph of the Fourier series for Example 1.29

Step 5'.   To show Theorem 1.28, we need to establish

$$\int_{-\pi}^{\pi} f(u+x) P_N(u)\, du \quad \to \quad \frac{f(x+0) + f(x-0)}{2} \tag{1.29}$$

as $N \to \infty$.

We show (1.29) by establishing the following two limits:

$$\int_{0}^{\pi} f(u+x) P_N(u)\, du \quad \to \quad \frac{f(x+0)}{2}$$

$$\int_{-\pi}^{0} f(u+x) P_N(u)\, du \quad \to \quad \frac{f(x-0)}{2}.$$

Using Step 4', these limits are equivalent to the following two limits:

$$\int_{0}^{\pi} (f(u+x) - f(x+0)) P_N(u)\, du \quad \to \quad 0 \tag{1.30}$$

$$\int_{-\pi}^{0} (f(u+x) - f(x-0)) P_N(u)\, du \quad \to \quad 0. \tag{1.31}$$

Using the definition of $P_N(u)$, equation (1.30) is equivalent to showing

$$\frac{1}{2\pi} \int_{0}^{\pi} \left(\frac{f(x+u) - f(x+0)}{\sin(u/2)} \right) \sin((N+1/2)u)\, du \quad \to \quad 0.$$

This limit follows from the Riemann-Lebesgue lemma exactly as in Step 5. Since u is positive in the preceding integral, we only need to know that the expression in parentheses is continuous from the right (i.e., has a right limit as $u \to 0^+$). Since f is assumed to be right differentiable, the same limit argument from Step 5 can be repeated here (with $u > 0$).

Similar arguments can be used to establish (1.31). This completes the proof of Theorem 1.28. ♦

1.3.4 Uniform Convergence

Now we discuss uniform convergence of Fourier series. As stated in Definition 0.8, a sequence of functions $F_n(x)$ converges *uniformly* to $F(x)$ if the rate of convergence is independent of the point x. In other words, given any small tolerance, $\epsilon > 0$ (such as $\epsilon = .01$), there exists a number N that is independent of x, such that $|F_n(x) - F(x)| < \epsilon$ for all x and all $n \geq N$. If the F_n did not converge uniformly, then we might have to choose different values of N for different x values to achieve the same degree of accuracy.

We say that the *Fourier series of $f(x)$ converges to $f(x)$ uniformly* if the sequence of partial sums

$$S_N(x) = a_0 + \sum_{k=1}^{N} a_k \cos(kx) + b_k \sin(kx)$$

converges to $f(x)$ uniformly as $N \to \infty$, where the a_k and b_k are the Fourier coefficients of f.

From Figures 8 and 9, it appears that the Fourier series in Example 1.10 converges uniformly. By contrast, the Fourier series in Example 1.9 does not appear to converge uniformly to $f(x)$. As the point x gets closer to a point of discontinuity of f, the rate of convergence gets slower. In view of Example 1.9, the number of terms, N, that must be used in the partial sum of the Fourier series to achieve a certain degree of accuracy must increase as x approaches a point of discontinuity.

In order to state the following theorem on uniform convergence, one more definition is needed. A function is said to be *piecewise smooth* if it is continuous and its derivative is defined everywhere except possibly for a discrete set of points. For example, the sawtooth function in Example 1.10 is piecewise smooth since the derivative of f exists at all points except at multiples of $\pi/2$ (which is a discrete set of points).

We now state the theorem for uniform convergence of Fourier series for the interval $[-\pi, \pi]$. This theorem also holds with π replaced by any number a.

THEOREM 1.30 *The Fourier series of a piecewise smooth, 2π-periodic function $f(x)$ converges uniformly to $f(x)$ on $[-\pi, \pi]$.*

EXAMPLE 1.31

Consider the sawtooth wave in Example 1.10. The graph of its periodic extension is piecewise smooth, as is clear from Figure 7. Therefore, Theorem 1.30 guarantees that its Fourier series converges uniformly.   ∎

EXAMPLE 1.32

Consider the Fourier series for the function $f(x) = x$ on the interval $[-\pi, \pi]$ considered in Example 1.9. Since $f(x) = x$ is not periodic, we need to consider its periodic extension, which is graphed in Figure 4. Note that its periodic extension is not continuous and therefore not piecewise smooth [even though $f(x) = x$ is everywhere smooth]. Therefore, Theorem 1.30 does not apply. In fact, due to the Gibbs effect in this example (see Figures 5 and 6), the Fourier series for this example does not converge uniformly.   ∎

Proof of Theorem 1.30 We prove this theorem under the simplifying assumption that f is everywhere twice differentiable.

As a first step, we prove the following relationship between the Fourier coefficients of f and the corresponding Fourier coefficients of f'': If

$$f(x) = \sum_n a_n \cos(nx) + b_n \sin(nx) \quad \text{and} \quad f''(x) = \sum_n a_n'' \cos(nx) + b_n'' \sin(nx)$$

then

$$a_n = \frac{-a_n''}{n^2} \tag{1.32}$$

$$b_n = \frac{-b_n''}{n^2}. \tag{1.33}$$

To establish the first relation, we use integration by parts on the integral formula for a_n (Theorem 1.2) to obtain

$$a_n = \frac{1}{\pi} \int_{-\pi}^{\pi} f(x) \cos(nx) \, dx$$

$$= f(x) \frac{\sin(nx)}{n} \Big|_{-\pi}^{\pi} - \frac{1}{\pi} \int_{-\pi}^{\pi} f'(x) \frac{\sin(nx)}{n} \, dx.$$

The first term on the right is zero since $\sin n\pi = \sin(-n\pi) = 0$. The second term on the right is $-b_n'/n$, where b_n' is the Fourier sine coefficient of f'. Repeating the same process [this time with $dv = (\sin nx)/n$ and $u = f'$] gives

$$a_n = \frac{-1}{\pi n^2} \int_{-\pi}^{\pi} f''(x) \cos(nx) \, dx.$$

The right side is $-a_n''/n^2$, as claimed in (1.32). Equation (1.33) for b_n is proved in a similar manner.

If f'' is continuous, then both the a_n'' and b_n'' stay bounded by some number M (in fact, by the Riemann Lebesgue lemma, a_n'' and b_n'' converge to zero as $n \to \infty$). Therefore, using (1.32) and (1.33),

$$\sum_{n=1}^{\infty} |a_n| + |b_n| = \sum_{n=1}^{\infty} \frac{|a_n''| + |b_n''|}{n^2} \leq \sum_{n=1}^{\infty} \frac{M + M}{n^2}.$$

This series is finite by the integral test for series (i.e., $\sum_{n=1}^{\infty} 1/n^2$ is finite since $\int_1^{\infty} dx/x^2$ is finite). The proof of the theorem will therefore follow from the following lemma.

LEMMA 1.33   *Suppose*

$$f(x) = a_0 + \sum_{k=1}^{\infty} a_k \cos(kx) + b_k \sin(kx)$$

with

$$\sum_{k=1}^{\infty} |a_k| + |b_k| < \infty.$$

Then the Fourier series converges uniformly and absolutely to the function $f(x)$.

Proof   We start with the estimate

$$|a_k \cos(kx) + b_k \sin(kx)| \leq |a_k| + |b_k| \tag{1.34}$$

(valid since $|\cos t|, |\sin t| \leq 1$). Thus the rate of convergence of the Fourier series of f at any point x is governed by the rate of convergence of $\sum_k |a_k| + |b_k|$. More precisely, let

$$S_N(x) = a_0 + \sum_{k=1}^{N} a_k \cos(kx) + b_k \sin(kx).$$

Then

$$f(x) - S_N(x) = a_0 + \sum_{k=1}^{\infty} a_k \cos(kx) + b_k \sin(kx)$$
$$- \left(a_0 + \sum_{k=1}^{N} a_k \cos(kx) + b_k \sin(kx) \right).$$

The a_0 and the terms up through $k = N$ cancel. Thus

$$f(x) - S_N(x) = \sum_{k=N+1}^{\infty} a_k \cos(kx) + b_k \sin(kx).$$

By (1.34)

$$|f(x) - S_N(x)| \le \sum_{k=N+1}^{\infty} |a_k| + |b_k| \qquad (1.35)$$

uniformly for all x. Since the series $\sum_{k=1}^{\infty} |a_k| + |b_k|$ converges by hypothesis, the tail end of this series can be made as small as desired by choosing N large enough. So, given $\epsilon > 0$, there is an integer $N_0 > 0$ so that if $N > N_0$, then $\sum_{k=N+1}^{\infty} |a_k| + |b_k| < \epsilon$. From (1.35)

$$|f(x) - S_N(x)| < \epsilon \quad \text{for} \quad N > N_0$$

for all x. N does not depend on x; N depends only on the rate of convergence of $\sum_{k=1}^{\infty} |a_k| + |b_k|$. Therefore, the convergence of $S_N(x)$ is uniform. This completes the proof of Lemma 1.33 and of Theorem 1.30. ◆

1.3.5 Convergence in the Mean

As pointed out in the previous section, if f is not continuous, then its Fourier series does not converge to $f(x)$ at points where $f(x)$ is discontinuous (it converges to the average of its left and right limits instead). In cases where a Fourier series does not converge uniformly or pointwise, it may converge in a weaker sense, such as in L^2 (in the mean). We investigate L^2 convergence of Fourier series in this section. Again, we state and prove the results in this section for 2π-periodic functions. However, the results remain true for other intervals as well (by replacing π by any number a and by using the appropriate form of the Fourier series for the interval $[-a, a]$).

First, we recall some concepts from Chapter 0 on inner product spaces. We will be working with the space $V = L^2([-\pi, \pi])$ consisting of all square integrable functions (i.e., f with $\int_{-\pi}^{\pi} |f(x)|^2 \, dx < \infty$). V is an inner product space with the following inner product:

$$\langle f, g \rangle = \int_{-\pi}^{\pi} f(x)\overline{g(x)} \, dx.$$

The norm $\|f\|$ in this space is therefore defined by

$$\|f\|^2 = \int_{-\pi}^{\pi} |f(x)|^2 \, dx.$$

We remind you of the two most important inequalities of an inner product space:

$$\langle f, g \rangle_V \le \|f\| \, \|g\| \qquad \text{and} \qquad \|f + g\| \le \|f\| + \|g\|.$$

The first of these is the *Schwarz inequality* and the second is the *triangle inequality*.

Let

$$V_N = \text{the space spanned by } \{1, \cos(kx), \sin(kx), k = 1, \ldots, N\}.$$

An element in V_N is a sum of the form

$$c_0 + \sum_{k=1}^{N} c_k \cos(kx) + d_k \sin(kx)$$

where c_k and d_k are any complex numbers. Suppose f belongs to $L^2[-\pi, \pi]$.
Let

$$f_N(x) = a_0 + \sum_{k=1}^{N} a_k \cos(kx) + b_k \sin(kx) \in V_N$$

be its partial Fourier series, where the a_k and b_k are the Fourier coefficients
given in Theorem 1.2. The key point in the proof of Theorem 1.2 is that a_k
and b_k are obtained by orthogonally projecting f onto the space spanned by
$\cos(kx)$ and $\sin(kx)$ (see the remark at the end of the proof). Thus, f_N is the
orthogonal projection of f onto the space V_N. In particular, f_N is the element
in V_N that is closest to f in the L^2 sense. We summarize this discussion in the
following lemma.

LEMMA 1.34   *Suppose f is an element of $V = L^2([-\pi, \pi])$. Let V_N be the
linear span of $\{1, \cos(kx), \sin(kx), 1 \leq k \leq N\}$. Let*

$$f_N(x) = a_0 + \sum_{k=1}^{N} a_k \cos(kx) + b_k \sin(kx)$$

*where a_k and b_k are the Fourier coefficients of f. Then f_N is the element in
V_N which is the closest to f in the L^2-norm; that is,*

$$\|f - f_N\|_{L^2} = \min_{g \in V_N} \|f - g\|_{L^2}. \qquad \blacklozenge$$

The main result of this section is contained in the next theorem.

THEOREM 1.35   *Suppose f is an element of $L^2([-\pi, \pi])$. Let*

$$f_N(x) = a_0 + \sum_{k=1}^{N} a_k \cos(kx) + b_k \sin(kx)$$

*where a_k and b_k are the Fourier coefficients of f. Then f_N converges to f in
$L^2([-\pi, \pi])$; that is, $\|f_N - f\|_{L^2} \to 0$ as $N \to \infty$.*

Theorem 1.35 also holds for the complex form of Fourier series.

THEOREM 1.36 *Suppose f is an element of $L^2([-\pi, \pi])$ with (complex) Fourier coefficients given by*

$$\alpha_n = \frac{1}{2\pi} \int_{-\pi}^{\pi} f(t) e^{-int} \, dt \quad \text{for } n \in Z.$$

Then the partial sum

$$f_N(t) = \sum_{k=-N}^{N} \alpha_k e^{ikt}$$

converges to f in the $L^2([-\pi, \pi])$ norm as $N \to \infty$.

EXAMPLE 1.37

All the examples considered in this chapter arise from functions that are in L^2 (over the appropriate interval under consideration for each example). Therefore, the Fourier series of each example in this chapter converges in the mean. ∎

Proof The proofs of both theorems are very similar. We give the proof of Theorem 1.35.

The proof involves two key steps. The first step (the next lemma) states that any function in $L^2([-\pi, \pi])$ can be approximated in the L^2-norm by a piecewise smooth periodic function g. The second step (Theorem 1.30) is to approximate g uniformly (and therefore in L^2) by its Fourier series. We start with the following lemma.

LEMMA 1.38 *A function in $L^2([-\pi, \pi])$ can be approximated arbitrarily closely by a smooth, 2π-periodic function.*

A rigorous proof of this lemma is beyond the scope of this book. However, we can give an intuitive idea as to why this lemma holds. A typical element $f \in L^2[-\pi, \pi]$ is not continuous. Even if it were continuous, its periodic extension is often not continuous. The idea is to connect the continuous components of f with the graph of a smooth function g. This is illustrated in Figures 18 through 20. In Figure 18, the graph of a typical $f \in L^2[-\pi, \pi]$ is given. The graph of its periodic extension is given in Figure 19. In Figure 20, the graph of a continuous g that connects the continuous components of f is superimposed on the graph of f. Rounding the corners of the connecting segments then molds g into a smooth function. Since the extended f is periodic, we can arrange that g is periodic as well.

The graph of g agrees with the graph of f everywhere except on the connecting segments that connect the continuous components of f. Since the horizontal width of each of these segments can be made very small (by increasing the slopes of these connecting segments), g can be chosen very close to f in the L^2-norm. These ideas are explored in more detail in Exercise 27.

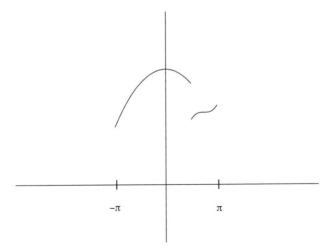

Figure 18 Typical f in L^2

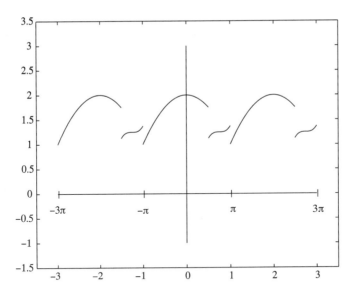

Figure 19 Periodic extension of f

Now we can complete the proof of Theorem 1.35. Using Lemma 1.38 and Theorem 0.10, we can (for any $\epsilon > 0$) choose a differentiable, periodic function, g, with

$$\|f - g\|_{L^2} < \epsilon. \tag{1.36}$$

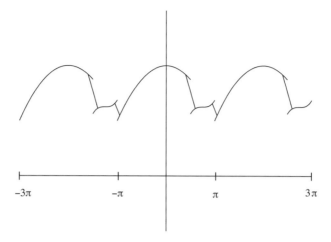

Figure 20 Approximation of f by a continuous g

Let

$$g_N(x) = c_0 + \sum_{k=1}^{N} c_k \cos(kx) + d_k \sin(kx)$$

where c_k and d_k are the Fourier cosine and sine coefficients for g. Since g is differentiable and periodic, we can uniformly approximate g by g_N (using Theorem 1.30). By choosing N_0 large enough, we can arrange $|g(x) - g_N(x)| < \epsilon$ for all $x \in [-\pi, \pi]$ and for $N > N_0$. Therefore,

$$\|g - g_N\|^2 = \int_{-\pi}^{\pi} |g(x) - g_N(x)|^2 \, dx \leq \int_{-\pi}^{\pi} \epsilon^2 \, dx \quad \text{if } N > N_0$$
$$= 2\pi\epsilon^2.$$

By taking square roots,

$$\|g - g_N\| < \sqrt{2\pi}\epsilon.$$

Combining this estimate with (1.36), we obtain

$$\|f - g_N\| = \|f - g + g - g_N\|$$
$$\leq \|f - g\| + \|g - g_N\| \quad \text{(triangle inequality)}$$
$$< \epsilon + \sqrt{2\pi}\epsilon \quad \text{for } N > N_0.$$

Now g_N is a linear combination of $\sin(kx)$ and $\cos(kx)$ for $k \leq N$ and therefore g_N belongs to V_N. We have already shown that f_N is the closest element from V_N to f in the L^2-norm (see Lemma 1.34). Therefore, we conclude that

$$\|f - f_N\| \leq \|f - g_N\| < (1 + \sqrt{2\pi})\epsilon \quad \text{for } N > N_0.$$

Since the tolerance, ϵ, can be chosen as small as desired, the proof of Theorem 1.35 is complete.   ◆

One consequence of Theorems 1.35 and 1.36 is the following theorem which is known as *Parseval's equation*. We state both the real and complex versions.

THEOREM 1.39   Parseval's Equation—Real Version. *Suppose*

$$f(x) = a_0 + \sum_{k=1}^{\infty} a_k \cos(kx) + b_k \sin(kx) \ \in \ L^2[-\pi, \pi].$$

Then

$$\frac{1}{\pi} \int_{-\pi}^{\pi} |f(x)|^2 \, dx = 2|a_0|^2 + \sum_{k=1}^{\infty} |a_k|^2 + |b_k|^2. \tag{1.37}$$

THEOREM 1.40   Parseval's Equation—Complex Version. *Suppose*

$$f(x) = \sum_{k=-\infty}^{\infty} \alpha_k e^{ikx} \ \in \ L^2[-\pi, \pi].$$

Then

$$\frac{1}{2\pi} \|f\|^2 = \frac{1}{2\pi} \int_{-\pi}^{\pi} |f(x)|^2 \, dx = \sum_{k=-\infty}^{\infty} |\alpha_k|^2. \tag{1.38}$$

Moreover, for f and g in $L^2[-\pi, \pi]$,

$$\frac{1}{2\pi} \langle f, g \rangle = \frac{1}{2\pi} \int_{-\pi}^{\pi} f(t)\overline{g(t)} \, dt = \sum_{n=-\infty}^{\infty} \alpha_n \overline{\beta_n}. \tag{1.39}$$

Remark.   The L^2-norm of a signal is often interpreted as its energy. With this physical interpretation, the squares of the Fourier coefficients of a signal measure the energy of the corresponding frequency components. Therefore, a physical interpretation of Parseval's equation is that the energy of a signal is simply the sum of the energies from each of its frequency components. (See Example 1.41.)

Proof   We prove the complex version of Parseval's equation. The proof of the real version is similar.

We prove equation (1.39). Equation (1.38) then follows from equation (1.39) by setting $f = g$.

Let

$$f_N(x) = \sum_{k=-N}^{N} \alpha_k e^{ikx}$$

$$g_N(x) = \sum_{k=-N}^{N} \beta_k e^{ikx}$$

be the partial sum of the Fourier series of f and g, respectively. By Theorem 1.36 , $f_N \to f$ and $g_N \to g$ in $L^2[-\pi,\pi]$ as $N \to \infty$. We have

$$\langle f_N, g_N \rangle = \langle \sum_{k=-N}^{N} \alpha_k e^{ikx}, \sum_{n=-N}^{N} \beta_n e^{inx} \rangle = \sum_{k=-N}^{N} \sum_{n=-N}^{N} \alpha_k \overline{\beta_n} \langle e^{ikx}, e^{inx} \rangle.$$

Since $\{e^{ikx}/\sqrt{2\pi}, k = \ldots, -1, 0, 1, \ldots\}$ is orthonormal, $\langle e^{ikx}, e^{inx} \rangle$ is 0 if $k \neq n$ and 2π if $k = n$. Therefore,

$$\langle f_N, g_N \rangle = \sum_{n=-N}^{N} \alpha_n \overline{\beta_n} \langle e^{inx}, e^{inx} \rangle = 2\pi \sum_{n=-N}^{N} \alpha_n \overline{\beta_n}.$$

Equation (1.39) follows by letting $N \to \infty$ provided that we show

$$\langle f_N, g_N \rangle \to \langle f, g \rangle \quad \text{as} \quad N \to \infty. \tag{1.40}$$

To show (1.40), we write

$$|\langle f, g, \rangle - \langle f_N, g_N \rangle| = |(\langle f, g, \rangle - \langle f, g_N \rangle) + (\langle f, g_N \rangle - \langle f_N, g_N \rangle)|$$
$$\leq |\langle f, g - g_N \rangle| + |\langle f - f_N, g_N \rangle|$$
$$\leq \|f\| \|g - g_N\| + \|f - f_N\| \|g_N\|$$

where the last step follows by Schwarz's inequality. Since $\|f_N - f\| \to 0$ and $\|g - g_N\| \to 0$ in L^2 (Theorem 1.36), the right side converges to zero as $N \to \infty$ and (1.40) follows. ♦

Note. If the series on the right side of (1.38) is truncated at some finite value N, then the right side can only get smaller, resulting in the following inequality:

$$\sum_{k=-N}^{N} |\alpha_k|^2 \leq \frac{1}{2\pi} \|f\|^2.$$

This in known as *Bessel's inequality*.

EXAMPLE 1.41

As we noted earlier, one interpretation of Parseval's theorem is that the energy in a signal is the sum of the energies associated with its Fourier components. In Example 1.9, we found that the Fourier coefficients for $f(x) = x$, $-\pi \leq x < \pi$ are $a_n = 0$ and $b_n = 2(-1)^{n+1}/n$. By equation (1.37), we have

$$\frac{1}{\pi} \int_{-\pi}^{\pi} x^2 \, dx = \sum_{n=1}^{\infty} \frac{4}{n^2}.$$

Evaluating the integral on the left and dividing both sides by 4, we see that

$$\sum_{n=1}^{\infty} \frac{1}{n^2} = \frac{\pi^2}{6}.$$

This sum is computed another way in Exercise 18.   ■

1.4 Exercises

1. Expand the function $f(x) = x^2$ in a Fourier series valid on the interval $-\pi \leq x \leq \pi$. Plot both f and the partial sums of its Fourier series

$$S_N(x) = \sum_{k=0}^{N} a_k \cos(kx) + b_k \sin(kx)$$

   for $N = 1, 2, 5, 7$. Observe how the graphs of the partial sums $S_N(x)$ approximate the graph of f. Plot the same graphs over the interval $-2\pi \leq x \leq 2\pi$.

2. Repeat the previous exercise for the interval $-1 \leq x \leq 1$. That is, expand the function $f(x) = x^2$ in a Fourier series valid on the interval $-1 \leq x \leq 1$. Plot both f and the partial Fourier series

$$\sum_{k=0}^{N} a_k \cos(\pi kx) + b_k \sin(\pi kx)$$

   for $N = 1, 2, 5, 7$ over the interval $-1 \leq x \leq 1$ and $-2 \leq x \leq 2$.

3. Expand the function $f(x) = x^2$ in a Fourier cosine series on the interval $0 \leq x \leq \pi$.

4. Expand the function $f(x) = x^2$ in a Fourier sine series on the interval $0 \leq x \leq 1$.

5. Expand the function $f(x) = x^3$ in a Fourier cosine series on the interval $0 \leq x \leq \pi$.

6. Expand the function $f(x) = x^3$ in a Fourier sine series on the interval $0 \leq x \leq 1$.

7. Expand the function $f(x) = |\sin x|$ in a Fourier series valid on the interval $-\pi \leq x \leq \pi$.

8. Expand the following function:

$$f(x) = \begin{cases} 1 & -1/2 < x \leq 1/2 \\ 0 & -1 < x \leq -1/2 \text{ or } 1/2 < x \leq 1 \end{cases}$$

in a Fourier series valid on the interval $-1 \le x \le 1$. Plot the various partial Fourier series along with the graph of f as in problem 1 for $N = 5,\ 10,\ 20$, and 40 terms. Notice how much slower the series converges to f in this example than in Exercise 1. What accounts for the slow rate of convergence in this example?

9. Expand the function $f(x) = e^{rx}$ in a Fourier series valid for $-\pi \le x \le \pi$. For the case $r = 1/2$, plot the partial Fourier series along with the graph of f as in problem 1 for $N = 10, 20$ and 30 terms. Plot these functions over the interval $-\pi \le x \le \pi$ and also $-2\pi \le x \le 2\pi$.

10. Use the previous problem to compute the Fourier coefficients for the function $f(x) = \sinh x = (e^x - e^{-x})/2$ and $f(x) = \cosh(x) = (e^x + e^{-x})/2$ over the interval $-\pi \le x \le \pi$.

11. Expand the function $f(x) = \cos x$ in a Fourier sine series on the interval $0 \le x \le \pi$.

12. Show that

$$\int_0^1 \cos(2n\pi x) \sin(2k\pi x)\, dx = 0.$$

13. Show that

$$\left\{\ldots, \frac{1}{\sqrt{a}}\cos\left(\frac{2\pi t}{a}\right), \frac{1}{\sqrt{a}}\cos\left(\frac{\pi t}{a}\right), \frac{1}{\sqrt{2a}}, \frac{1}{\sqrt{a}}\sin\left(\frac{\pi t}{a}\right), \frac{1}{\sqrt{a}}\sin\left(\frac{2\pi t}{a}\right), \ldots\right\}$$

is an orthonormal set of functions on $L^2([-a, a])$. Establish the proof of Theorem 1.4.

14. Prove Lemma 1.16 and Theorem 1.20.

15. Let $F(x)$ be the Fourier series for the function

$$f(x) = \begin{cases} 1 & -1/2 < x \le 1/2 \\ 0 & -1 < x \le -1/2 \text{ or } 1/2 < x \le 1. \end{cases}$$

State the precise value of $F(x)$ for each x in the interval $-1 \le x \le 1$.

16. If $f(x)$ is continuous on the interval $0 \le x \le a$, show that its even periodic extension is continuous everywhere. Does this statement hold for the odd periodic extension? What additional condition(s) is (are) necessary to ensure that the odd periodic extension is everywhere continuous?

17. Consider the sawtooth function and the Fourier series derived for it Example 1.10.

   (a) Use the convergence theorems in this chapter to explain why the Fourier series for the sawtooth function converges pointwise to that function.

(b) Use this fact to show that

$$\sum_{k=0}^{\infty} \frac{1}{(2k+1)^2} = \frac{\pi^2}{8}.$$

18. In Exercise 1, you found the Fourier series for the function $f(x) = x^2$, $-\pi \leq x \leq \pi$. Explain why this series converges uniformly to x^2 on $[-\pi, \pi]$. Use this to show that $\sum_{n=1}^{\infty} \frac{1}{n^2} = \frac{\pi^2}{6}$. (*Hint:* What happens at $x = \pi$?)

19. Sketch two periods of the pointwise limit of the Fourier series for each of the following functions. State whether or not each function's Fourier series converges uniformly. (You do not need to compute the Fourier coefficients.)

   (a) $f(x) = e^x$, $-1 < x \leq 1$

   (b) $f(x) = \begin{cases} 1 & -1/2 < x \leq 1/2 \\ 0 & -1 < x \leq -1/2 \text{ or } 1/2 < x \leq 1 \end{cases}$

   (c) $f(x) = x - x^2$, $-1 < x \leq 1$

   (d) $f(x) = 1 - x^2$, $-1 < x \leq 1$

   (e) $f(x) = \cos x + |\cos x|$, $-\pi < x \leq \pi$

   (f) $f(x) = \begin{cases} \frac{\sin x}{x} & -\pi < x \leq \pi, \ x \neq 0 \\ 1 & x = 0 \end{cases}$

20. For each of the functions in Exercise 19, state whether or not its Fourier sine and cosine series (for the corresponding half-interval) converge uniformly on the entire real line, $-\infty < x < \infty$.

21. If F is 2π-periodic and c is any real number, then show that

$$\int_{-\pi}^{-\pi+c} F(x)\,dx = \int_{\pi}^{\pi+c} F(x)\,dx.$$

   Hint: Use the change of variables $x = t - 2\pi$. Then use this equation to prove Lemma 1.3.

22. If f is a real-valued even function on the interval $[-\pi, \pi]$, show that the complex Fourier coefficients are real. If f is a real-valued odd function on the interval $[-\pi, \pi]$, show that the complex Fourier coefficients are purely imaginary (i.e., their real parts are zero).

23. Suppose f is continuously differentiable [i.e., $f'(x)$ is continuous for all x] and 2π-periodic. Without quoting Theorem 1.35, show that the Fourier series of f converges in the mean. *Hint:* Use the relationship between the Fourier coefficients of f and those of f' given in the proof of Theorem 1.30.

24. From Theorem 1.8, the Fourier series of an odd function consists only of sine terms. What additional symmetry conditions on f will imply that the sine coefficients with even indices will be zero? Give an example of a function satisfying this additional condition.

25. Suppose that there is a sequence of nonnegative numbers $\{M_n\}$ such that $\sum_n M_n$ is convergent. Suppose a_n and b_n are given with

$$|a_n|, \ |b_n| \leq M_n \quad \text{for all} \quad n \geq 0.$$

Show that the following series is uniformly convergent on $-\infty < x < \infty$.

$$a_0 + \sum_{n=1}^{\infty} a_n \cos nx + b_n \sin nx$$

26. Prove the following version of the Riemann-Lebesgue lemma for infinite intervals: Suppose f is a continuous function on $a \leq t < \infty$ with $\int_a^\infty |f(t)| \, dt < \infty$; show that

$$\lim_{n \to \infty} \int_a^\infty f(t) \cos nt \, dt = \lim_{n \to \infty} \int_a^\infty f(t) \sin nt \, dt = 0.$$

Hint: Break up the interval $a \leq t < \infty$ into two intervals, $a \leq t \leq M$ and $M \leq t < \infty$, where M is chosen so large that $\int_M^\infty |f(t)| \, dt$ is less than $\epsilon/2$; apply the usual Riemann-Lebesgue lemma to the first interval.

27. This exercise explores the ideas in the proof of Lemma 1.38 with a specific function. Let

$$f(t) = \begin{cases} 0 & 0 \leq t \leq 1/2 \\ 1 & 1/2 < t \leq 1. \end{cases}$$

For $0 < \delta < 1/2$, let

$$g_\delta(t) = \begin{cases} 0 & 0 \leq t \leq 1/2 - \delta \\ \frac{t}{2\delta} - \frac{1}{4\delta} + \frac{1}{2} & 1/2 - \delta < t \leq 1/2 + \delta \\ 1 & 1/2 + \delta < t \leq 1. \end{cases}$$

Graph both f and g_δ for $\delta = 0.1$. Show that as $\delta \mapsto 0$, $||f - g_\delta||_{L^2[0,1]} \mapsto 0$. Note that g_δ is continuous but not differentiable. Can you modify the formula for g_δ so that g_δ is C^1 (one continuous derivative) and still satisfy $\lim_{\delta \to 0} ||f - g_\delta||_{L^2[0,1]} = 0$?

28. This exercise explains the Gibbs phenomenon, which is evident in the convergence of Fourier series near a point of discontinuity. We examine the Gibbs phenomenon for the Fourier series of the function

$$f(t) = \begin{cases} \pi - t & 0 \leq t \leq \pi \\ -\pi - t & -\pi \leq t < 0 \end{cases}$$

on the interval $-\pi \leq t \leq \pi$ (see Figure 21 for the graph of f and its partial Fourier series). Complete the following outline.

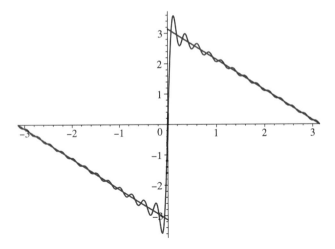

Figure 21 Gibbs phenomenon

(a) Show that the Fourier series for f is

$$2\sum_{n=1}^{\infty}\frac{\sin nx}{n}.$$

(b) Let

$$g_N(x) = 2\sum_{n=1}^{N}\frac{\sin nx}{n} - (\pi - x)$$

where g_N represents the difference between the $f(x)$ and the Nth partial sum of the Fourier series of f. In the remaining parts of this exercise, you will show that the maximum value of g_N is greater than 0.5 when N is large and thus represents the Gibbs effect.

(c) Show that $g'_N(x) = 2\pi P_N(x)$, where

$$P_N(x) = \frac{1}{\pi}\left(\frac{1}{2} + \cos x \cdots + \cos Nx\right)$$

$$= \frac{1}{2\pi}\frac{\sin(N+1/2)x}{\sin(x/2)} \quad \text{by Lemma 1.23.}$$

(d) Show that $\theta_N = \pi/(N+1/2)$ is the first critical point of g_N to the right of $x = 0$.

(e) Use the fundamental theorem of calculus and part (c) to show

$$g_N(\theta_N) = \int_0^{\theta_N}\frac{\sin(N+1/2)x}{\sin(x/2)}\,dx - \pi.$$

(f) Show

$$\lim_{N\to\infty} g_N(\theta_N) = 2\int_0^\pi \frac{\sin x}{x}\,dx - \pi.$$

Hint: Make a change of variables $\phi = (N+1/2)x$ and use the fact that $\sin t/t \to 1$ as $t \to 0$.

(g) Show that

$$2\int_0^\pi \frac{\sin x}{x}\,dx - \pi \approx 0.562$$

by evaluating this integral numerically. Thus the Gibbs effect or the amount that the Nth partial sum overshoots the function f is about 0.562 when N is large.

29. Use Parseval's theorem and the series from Exercise 1 to find the sum of the series $\sum_{n=1}^\infty \frac{1}{n^4}$.

The next two exercises require a computer algebra system (e.g., Maple or MAT-LAB) to compute Fourier coefficients.

30. Consider the function

$$f(x) = e^{-x^2/10}\left(\cos 2x + 2\sin 4x + 0.4\cos 2x \cos 40x\right).$$

For what values of n would you expect the Fourier coefficients $a(n)$ and $b(n)$ to be significant (say bigger than 0.01 in absolute value)? Why? Compute the $a(n)$ and $b(n)$ through $n = 50$ and see if you are right. Plot the partial Fourier series through $n = 6$ and compare with the plot of the original $f(x)$.

31. Consider the function

$$g(x) = e^{-x^2/8}\left(\cos 2x + 2\sin 4x + 0.4\cos 2x \cos 10x\right).$$

Compute the partial Fourier series through $N = 25$. Throw away any coefficients that are smaller than 0.01 in absolute value. Plot the resulting series and compare with the original function $g(x)$. Try experimenting with different tolerances (other than 0.01).

The remaining exercises pertain to Fourier series as a tool for solving partial differential equations, as indicated at the beginning of this chapter.

32. Solve the following heat equation problem:

$$u_t = u_{xx} \quad \text{for } t > 0,\ 0 \le x \le 1$$
$$u(0, x) = x - x^2 \quad \text{for } 0 \le x \le 1$$
$$u(0, t) = u(1, t) = 0.$$

33. If the boundary conditions $u(0,t) = A$ and $u(1,t) = B$ are not homogeneous (i.e., A and B are not necessarily zero), then the procedure given in Example 1.14 must be modified in order to solve the following heat equation:

$$u_t = u_{xx} \quad \text{for } t > 0,\ 0 \le x \le 1$$
$$u(x,t) = f(x) \quad \text{for } 0 \le x \le 1$$
$$u(0,t) = A \quad u(1,t) = B.$$

Let $L(x)$ be the linear function with $L(0) = A$ and $L(1) = B$ and let $\hat{u}(x,t) = u(x,t) - L(x)$. Show that $\hat{u}$ solves the following problem:

$$\hat{u}_t = \hat{u}_{xx} \quad \text{for } t > 0,\ 0 \le x \le 1$$
$$\hat{u}(x,t) = f(x) - L(x) \quad \text{for } 0 \le x \le 1$$
$$\hat{u}(0,t) = 0, \quad \hat{u}(1,t) = 0.$$

This heat equation can be solved for $\hat{u}$ using the techniques given in Example 1.14. The solution, u, to the original heat equation problem can then be found by the equation $u(x,t) = \hat{u}(x,t) + L(x)$.

34. Use the procedure outlined in the previous exercise to solve the following heat equation:

$$u_t = u_{xx} \quad \text{for } t > 0,\ 0 \le x \le 1$$
$$u(0,x) = 2 - x^2 \quad \text{for } 0 \le x \le 1$$
$$u(0,t) = 2, \quad u(1,t) = 1.$$

35. Another important version of the heat equation is the following Neumann boundary value problem:

$$u_t = u_{xx} \quad \text{for } t > 0,\ 0 \le x \le 1$$
$$u(0,x) = f(x) \quad \text{given for } 0 \le x \le 1$$
$$u_x(0,t) = 0, \quad u_x(1,t) = 0.$$

This problem represents the standard heat equation, where $u(x,t)$ is the temperature of a rod of unit length at position x and at time t; $f(x)$ is the initial (at time $t = 0$) temperature at position x. The boundary conditions $u_x = 0$ at $x = 0$ and $x = 1$ physically mean that no heat is escaping from the rod at its endpoints (i.e., the rod is insulated at its endpoints). Use the procedure outlined at the beginning of this chapter to show that the general solution to this problem is given by

$$\sum_{k=0}^{\infty} a_k e^{-(k\pi)^2 t} \cos(k\pi x)$$

where a_k are the coefficients of a Fourier cosine series of f over the interval $0 \le x \le 1$. Use this formula to obtain the solution in the case where $f(x) = x^2 - x$ for $0 \le x \le 1$.

36. The goal of this problem is to prove *Poisson's formula*, which states that if $f(t)$ is a piecewise smooth function on $-\pi \le t \le \pi$, then

$$u(r, \phi) = \frac{1}{2\pi} \int_0^{2\pi} f(t) \frac{1 - r^2 \, dt}{1 - 2\cos(\phi - t) + r^2}, \tag{1.41}$$

for $0 \le r \le 1$, $0 \le \phi \le 2\pi$ solves Laplaces equation

$$\Delta u = u_{xx} + u_{yy} = 0 \tag{1.42}$$

in the unit disc $x^2 + y^2 = 1$ (in polar coordinates: $\{r < 1\}$) with boundary values $u(r = 1, \phi) = f(\phi)$, $-\pi \le \phi \le \pi$. Follow the outline given to establish this formula:

(a) Show that the function $u(x, y) = (x + iy)^n$ solves Laplaces equation for each value of $n = 0, 1, 2, \ldots$ Using complex notation $z = x + iy$, this solution can be written as $u(z) = z^n$.

(b) Show that any finite sum of the form $\sum_{n=-N}^{N} A_n z^n$ solves Laplace's equation. It is a fact that if the infinite series (i.e., as $|N| \to \infty$) converges uniformly and absolutely for $|z| = 1$, then the infinite series $\sum_{n=-\infty}^{\infty} A_n z^n$ also solves Laplace's equation on $\{|z| < 1\}$. Write this function in polar coordinates with $z = re^{i\phi}$.

(c) In order to solve Laplace's equation, we must hunt for a solution of the form $u(r, \phi) = \sum_{n=-\infty}^{\infty} A_n r^{|n|} e^{in\phi}$ with boundary condition $u(r = 1, \phi) = f(\phi)$. Show that the boundary condition is satisfied if A_n is set to the Fourier coefficients of f in complex form.

(d) Using the formula for the complex Fourier coefficients, show that if f is real valued, then $A_{-n} = \overline{A_n}$. Use this fact to rewrite the solution in the previous step as

$$u(r, \phi) = \frac{1}{2\pi} \mathrm{Re} \left\{ \int_{-\pi}^{\pi} f(t) \left[2\left(\sum_{n=0}^{\infty} r^n e^{in(\phi - t)}\right) - 1 \right] \right\}.$$

(e) Now use the geometric series formula to rewrite the solution in the previous step as

$$u(r, \phi) = \frac{1}{2\pi} \int_{-\pi}^{\pi} f(t) P(r, \phi - t) \, dt$$

where

$$P(r, u) = \mathrm{Re}\left\{ \frac{2}{1 - re^{iu}} - 1 \right\}.$$

(f) Rewrite P as

$$P(r, u) = \frac{1 - r^2}{1 - 2r\cos u + r^2}.$$

Use this formula together with the previous integral formula for u to establish (1.41).

Chapter 2

The Fourier Transform

In this chapter, we develop the Fourier transform and its inverse. The Fourier transform can be thought of as a continuous form of Fourier series. A Fourier series decomposes a signal on $[-\pi, \pi]$ into components that vibrate at integer frequencies. By contrast, the Fourier transform decomposes a signal defined on an infinite time interval into a λ-frequency component, where λ can be any real (or even complex) number. Besides being of interest in their own right, the topics in this chapter will be important in the construction of wavelets in later chapters.

2.1 Informal Development of the Fourier Transform

2.1.1 The Fourier Inversion Theorem

In this section, we give an informal development of the Fourier transform and its inverse. Precise arguments are given in Appendix A.

To obtain the Fourier transform, we consider the Fourier series of a function defined on the interval $-l \leq x \leq l$ and then let l go to infinity. Recall from Theorem 1.20 (with $a = l$) that a function defined on $-l \leq x \leq l$ can be expressed as

$$f(x) = \sum_{n=-\infty}^{\infty} \alpha_n e^{in\pi x/l}$$

91

where

$$\alpha_n = \frac{1}{2l} \int_{-l}^{l} f(t)e^{-in\pi t/l}\, dt.$$

If f is defined on the entire real line, then it is tempting to let l go to infinity and see how the preceding formulas are affected. Substituting the expression for α_n into the previous sum, we obtain

$$f(x) = \lim_{l \to \infty} \left[\sum_{n=-\infty}^{\infty} \left(\frac{1}{2l} \int_{-l}^{l} f(t)e^{-in\pi t/l}\, dt \right) e^{in\pi x/l} \right]$$

$$= \lim_{l \to \infty} \left[\sum_{n=-\infty}^{\infty} \frac{1}{2l} \int_{-l}^{l} f(t)e^{in\pi(x-t)/l}\, dt \right].$$

Our goal is to recognize the sum on the right as the Riemann sum formulation of an integral. To this end, let $\lambda_n = \frac{n\pi}{l}$ and $\Delta\lambda = \lambda_{n+1} - \lambda_n = \frac{\pi}{l}$. We obtain

$$f(x) = \lim_{l \to \infty} \sum_{n=-\infty}^{\infty} \left[\frac{1}{2\pi} \int_{-l}^{l} f(t)e^{\lambda_n i(x-t)}\, dt \right] \Delta\lambda. \qquad (2.1)$$

Let

$$F_l(\lambda) = \frac{1}{2\pi} \int_{-l}^{l} f(t)e^{\lambda i(x-t)}\, dt.$$

The sum in (2.1) is now

$$\sum_{n=-\infty}^{\infty} F_l(\lambda_n)\Delta\lambda.$$

This term resembles the Riemann sum definition of the integral $\int_{-\infty}^{\infty} F_l(\lambda)\, d\lambda$. As l converges to ∞, the quantity $\Delta\lambda$ converges to 0 and so $\Delta\lambda$ becomes the $d\lambda$ in the integral $\int_{-\infty}^{\infty} F_l(\lambda)\, d\lambda$. So (2.1) becomes

$$f(x) = \lim_{l \to \infty} \int_{-\infty}^{\infty} F_l(\lambda)\, d\lambda.$$

As $l \mapsto \infty$, $F_l(\lambda)$ formally becomes the integral $\frac{1}{2\pi} \int_{-\infty}^{\infty} f(t)e^{i\lambda(x-t)}\, dt$. Therefore,

$$f(x) = \frac{1}{2\pi} \int_{-\infty}^{\infty} \int_{-\infty}^{\infty} f(t)e^{i\lambda(x-t)}\, dt\, d\lambda$$

or

$$f(x) = \frac{1}{\sqrt{2\pi}} \int_{-\infty}^{\infty} \left(\frac{1}{\sqrt{2\pi}} \int_{-\infty}^{\infty} f(t)e^{-i\lambda t}\, dt \right) e^{i\lambda x}\, dx. \qquad (2.2)$$

We let $\widehat{f}(\lambda)$ be the quantity inside the parentheses; that is,

$$\widehat{f}(\lambda) = \frac{1}{\sqrt{2\pi}} \int_{-\infty}^{\infty} f(t)e^{-i\lambda t}\, dt.$$

The function $\widehat{f}(\lambda)$ is called the *(complex form of the) Fourier transform of f.* Equation (2.2) becomes

$$f(x) = \frac{1}{\sqrt{2\pi}} \int_{-\infty}^{\infty} \widehat{f}(\lambda)e^{i\lambda x}\, d\lambda,$$

which is often referred to as the *Fourier inversion formula* since it describes $f(x)$ as an integral involving the Fourier transform of f.

We summarize this discussion in the following theorem.

THEOREM 2.1   *If f is a continuously differentiable function with $\int_{-\infty}^{\infty} |f(t)|\, dt < \infty$, then*

$$f(x) = \frac{1}{\sqrt{2\pi}} \int_{-\infty}^{\infty} \widehat{f}(\lambda)e^{i\lambda x}\, d\lambda \qquad (2.3)$$

where $\widehat{f}(\lambda)$ (the Fourier transform of f) is given by

$$\widehat{f}(\lambda) = \frac{1}{\sqrt{2\pi}} \int_{-\infty}^{\infty} f(t)e^{-i\lambda t}\, dt.$$

The preceding argument is not rigorous since we have not justified several steps, including the convergence of the improper integral $\int_{-\infty}^{\infty} F_l(\lambda)\, d\lambda$. As with the development of Fourier series (see Theorem 1.28), if the function $f(x)$ has points of discontinuity, such as a step function, then the preceding formula holds with $f(x)$ replaced by the average of the left- and right-hand limits [i.e., $(f(x+0) + f(x-0))/2$]. Rigorous proofs of these results are given in Appendix A.

The assumption $\int_{-\infty}^{\infty} |f(t)|\, dt < \infty$ in Theorem 2.1 is needed in order to make sense out of the improper integral defining $\widehat{f}$; that is,

$$|\widehat{f}(\lambda)| \le \frac{1}{\sqrt{2\pi}} \int_{-\infty}^{\infty} |f(t)e^{-i\lambda t}|\, dt$$

$$= \frac{1}{\sqrt{2\pi}} \int_{-\infty}^{\infty} |f(t)|\, dt \quad \text{since } |e^{-i\lambda t}| = 1$$

$$< \infty.$$

Comparison with Fourier Series.   The complex form of the Fourier transform of f and the corresponding inversion formula are analogous to the complex form of the Fourier series of f over the interval $-l \le x \le l$:

$$f(x) = \sum_{n=-\infty}^{\infty} \widehat{f}_n e^{\frac{in\pi}{l}x} \qquad (2.4)$$

where

$$\widehat{f}_n = \frac{1}{2l} \int_{-l}^{l} f(t) e^{\frac{-in\pi}{l} t} \, dt.$$

The variable λ in the Fourier inversion formula, (2.3), plays the role of $\frac{n\pi}{l}$ in (2.4). The sum over n from $-\infty$ to ∞ in (2.4) is replaced by an integral with respect to λ from $-\infty$ to ∞ defining (2.3). The formulas for $\widehat{f}_n$ and $\widehat{f}(\lambda)$ are also analogous. The integral over $[-l, \, l]$ in the formula for $\widehat{f}_n$ is analogous to the integral over $(-\infty, \, \infty)$ in $\hat{f}(\lambda)$. In the case of Fourier series, $\hat{f}_n$ measures the component of f that oscillates at frequency n. Likewise, $\hat{f}(\lambda)$ measures the frequency component of f that oscillates with frequency λ. If f is defined on a finite interval, then its Fourier series is a decomposition of f into a discrete set of oscillating frequencies (i.e., $\hat{f}_n$, one for each integer n). For a function on an infinite interval, there is a continuum of frequencies since the frequency component, $\hat{f}(\lambda)$, is defined for each real number λ. These ideas are illustrated in the following examples.

2.1.2 Examples

EXAMPLE 2.2
_____

In our first example, we will compute the Fourier transform of the rectangular wave (see Figure 1):

$$f(t) = \begin{cases} 1 & \text{if } -\pi \le t \le \pi \\ 0 & \text{otherwise.} \end{cases}$$

Now we have $f(t)e^{-i\lambda t} = f(t)(\cos \lambda t - i \sin \lambda t)$. Since f is an even function, $f(t)\sin(\lambda t)$ is an odd function and its integral over the real line is zero. Therefore, the Fourier transform of f is reduced to

$$\widehat{f}(\lambda) = \frac{1}{\sqrt{2\pi}} \int_{-\infty}^{\infty} f(t)\cos(\lambda t) \, dt$$

$$= \frac{1}{\sqrt{2\pi}} \int_{-\pi}^{\pi} \cos(\lambda t) \, dt$$

$$= \frac{\sqrt{2}\sin(\lambda \pi)}{\sqrt{\pi}\lambda}.$$

A graph of $\widehat{f}$ is given in Figure 2.

As already mentioned, the Fourier transform, $\widehat{f}(\lambda)$, measures the frequency component of f that vibrates with frequency λ. In this example, f is a piecewise constant function. Since a constant vibrates with zero frequency, we should expect that the largest values of $\widehat{f}(\lambda)$ occur when λ is near zero. The graph of $\widehat{f}$ clearly illustrates this feature. ∎

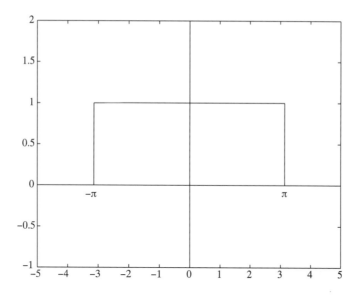

Figure 1 Rectangular wave

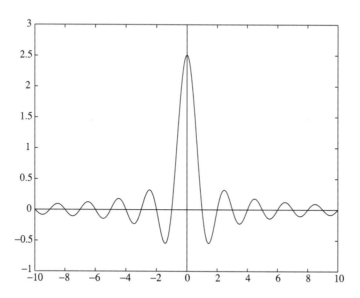

Figure 2 Fourier transform of a rectangular wave

EXAMPLE 2.3

Let
$$f(t) = \begin{cases} \cos 3t & \text{if } -\pi \le t \le \pi \\ 0 & \text{otherwise} \end{cases}$$

(see Figure 3).

Since f is an even function, only the cosine part of the transform contributes:

$$\widehat{f}(\lambda) = \frac{1}{\sqrt{2\pi}} \int_{-\infty}^{\infty} f(t)\cos(\lambda t)\,dt = \frac{1}{\sqrt{2\pi}} \int_{-\pi}^{\pi} \cos(3t)\cos(\lambda t)\,dt.$$

The preceding integral is left as an exercise [sum the identities

$$\cos(u + v) = \cos u \cos v - \sin u \sin v$$
$$\cos(u - v) = \cos u \cos v + \sin u \sin v$$

with $u = 3t$ and $v = \lambda t$ and then integrate]. The result is

$$\widehat{f}(\lambda) = \frac{\sqrt{2}\lambda \sin(\lambda \pi)}{\sqrt{\pi}(9 - \lambda^2)}.$$

The graph of $\widehat{f}$ is given in Figure 4.

Note that the Fourier transform peaks at $\lambda = 3$ and -3. This behavior should be expected since $f(t) = \cos(3t)$ vibrates with frequency 3 on the interval $-\pi \le t \le \pi$.   ■

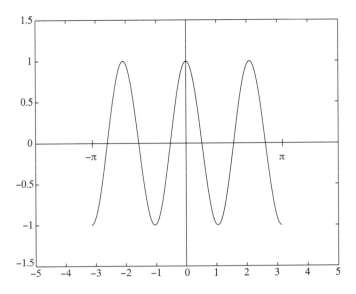

Figure 3 Plot of $\cos(3t)$

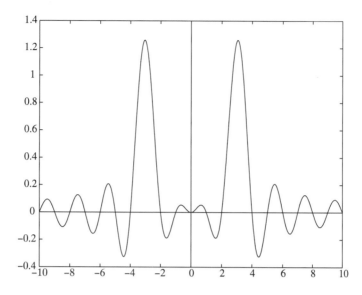

Figure 4 Fourier transform of $\cos(3t)$

EXAMPLE 2.4

Let

$$f(t) = \begin{cases} \sin 3t & \text{if } -\pi \leq t \leq \pi \\ 0 & \text{otherwise.} \end{cases}$$

Since f is an odd function, only the sine part of the transform contributes (which is purely imaginary). The transform is

$$\widehat{f}(\lambda) = \frac{-i}{\sqrt{2\pi}} \int_{-\infty}^{\infty} f(t)\sin(\lambda t)\, dt$$

$$= \frac{-i}{\sqrt{2\pi}} \int_{-\pi}^{\pi} \sin(3t)\sin(\lambda t)\, dt$$

$$= \frac{-3\sqrt{2}i\sin(\lambda\pi)}{\sqrt{\pi}(9 - \lambda^2)}.$$ ∎

EXAMPLE 2.5

The next example is a triangular wave whose graph is given in Figure 5. The analytical formula for this graph is given by

$$f(t) = \begin{cases} t + \pi & \text{if } -\pi \leq t \leq 0 \\ \pi - t & \text{if } 0 < t \leq \pi \\ 0 & \text{otherwise.} \end{cases}$$

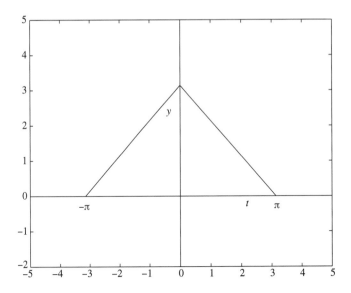

Figure 5 Triangular wave

This function is even and so its Fourier transform is given by

$$\widehat{f}(\lambda) = \frac{2}{\sqrt{2\pi}} \int_0^\infty f(t)\cos(\lambda t)\,dt = \frac{2}{\sqrt{2\pi}} \int_0^\pi (\pi - t)\cos(\lambda t)\,dt.$$

This integral can be computed using integration by parts:

$$\widehat{f}(\lambda) = \sqrt{\frac{2}{\pi}} \frac{(1 - \cos(\lambda\pi))}{\lambda^2}.$$

The graph of $\widehat{f}$ is given in Figure 6.

Note that the Fourier transforms in Examples 2.4 and 2.5 decay at the rate $1/\lambda^2$ as $\lambda \mapsto \infty$, which is faster than the decay rate of $1/\lambda$ exhibited by the Fourier transforms in Examples 2.2 and 2.3. The faster decay in Examples 2.4 and 2.5 results from the continuity of the functions in these examples. Note the parallel with the examples in Chapter 2. The Fourier coefficients, a_n and b_n, for the discontinuous function in Example 1.9 decay like $1/n$ whereas the Fourier coefficients for the continuous function in Example 1.10 decay like $1/n^2$.   ∎

2.2   Properties of the Fourier Transform

2.2.1   Basic Properties

In this section, we set down most of the basic properties of the Fourier transform. First, we introduce the alternative notation

$$\mathcal{F}[f](\lambda) = \widehat{f}(\lambda)$$

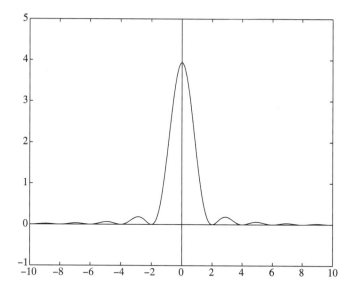

Figure 6 Fourier transform of the triangular wave

for the Fourier transform of f. This notation has advantages when discussing some of the operator theoretic properties of the Fourier transform. The Fourier operator $\mathcal{F}$ should be thought of as a mapping whose domain and range are the space of complex-valued functions defined on the real line. The input of $\mathcal{F}$ is a function, say f, and returns another function, $\mathcal{F}[f] = \widehat{f}$, as its output.

In a similar fashion, we define the inverse Fourier transform operator as

$$\mathcal{F}^{-1}[f](x) = \frac{1}{\sqrt{2\pi}} \int_{-\infty}^{\infty} f(\lambda)e^{i\lambda x}\, d\lambda.$$

Theorem 2.1 implies that $\mathcal{F}^{-1}$ really is the inverse of $\mathcal{F}$:

$$\mathcal{F}^{-1}[\mathcal{F}[f]] = f \qquad (2.5)$$

because

$$\mathcal{F}^{-1}[\mathcal{F}[f]](x) = \mathcal{F}^{-1}[\widehat{f}](x) \qquad \text{by definition of } \mathcal{F}$$
$$= \frac{1}{\sqrt{2\pi}} \int_{-\infty}^{\infty} \widehat{f}(\lambda)e^{i\lambda x}\, d\lambda \qquad \text{by definition of } \mathcal{F}^{-1}[\widehat{f}]$$
$$= f(x) \qquad \text{by Theorem 2.1.}$$

Some properties of the Fourier transform and its inverse are given in the following theorem. Other basic properties are given in the exercises.

THEOREM 2.6 *Let f and g be differentiable functions defined on the real line with $f(t) = 0$ for large $|t|$. The following properties hold.*

1. *The Fourier transform and its inverse are linear operators. That is, for any constant c*

$$\mathcal{F}[f+g] = \mathcal{F}[f] + \mathcal{F}[g] \quad and \quad \mathcal{F}[cf] = c\mathcal{F}[f].$$

$$\mathcal{F}^{-1}[f+g] = \mathcal{F}^{-1}[f] + \mathcal{F}^{-1}[g] \quad and \quad \mathcal{F}^{-1}[cf] = c\mathcal{F}^{-1}[f].$$

2. *The Fourier transform of a product of f with t^n is given by*

$$\mathcal{F}[t^n f(t)](\lambda) = i^n \frac{d^n}{d\lambda^n} \{\mathcal{F}[f](\lambda)\}.$$

3. *The inverse Fourier transform of a product of f with λ^n is given by*

$$\mathcal{F}^{-1}[\lambda^n f(\lambda)](t) = (-i)^n \frac{d^n}{dt^n} \{\mathcal{F}^{-1}[f](t)\}.$$

4. *The Fourier transform of an nth derivative is given by*

$$\mathcal{F}[f^{(n)}](\lambda) = (i\lambda)^n \mathcal{F}[f](\lambda)$$

[here $f^{(n)}$ stands for the nth derivative of f].

5. *The inverse Fourier transform of an nth derivative is given by*

$$\mathcal{F}^{-1}[f^{(n)}](t) = (-it)^n \mathcal{F}^{-1}[f](t).$$

6. *The Fourier transform of a translation is given by*

$$\mathcal{F}[f(t-a)](\lambda) = e^{-i\lambda a} \mathcal{F}[f](\lambda).$$

7. *The Fourier transform of a rescaling is given by*

$$\mathcal{F}[f(bt)](\lambda) = \frac{1}{b}\mathcal{F}[f](\frac{\lambda}{b}).$$

8. *If $f(t) = 0$ for $t < 0$, then*

$$\mathcal{F}[f](\lambda) = \frac{1}{\sqrt{2\pi}}\mathcal{L}[f](i\lambda)$$

where $\mathcal{L}[f]$ is the Laplace transform of f defined by

$$\mathcal{L}[f](s) = \int_0^\infty f(t)e^{-ts}\,dt.$$

Proof We prove each part separately.

1. The linearity of the Fourier transform follows from the linearity of the integral, as we demonstrate in the following:

$$\mathcal{F}[f + g](\lambda) = \frac{1}{\sqrt{2\pi}} \int_{-\infty}^{\infty} [f(t) + g(t)]e^{-i\lambda t}\, dt$$

$$= \frac{1}{\sqrt{2\pi}} \int_{-\infty}^{\infty} f(t)e^{-i\lambda t}\, dt + \frac{1}{\sqrt{2\pi}} \int_{-\infty}^{\infty} g(t)e^{-i\lambda t}\, dt$$

$$= \mathcal{F}[f](\lambda) + \mathcal{F}[g](\lambda).$$

The proof for $\mathcal{F}[cf] = c\mathcal{F}[f]$ is similar, as are the proofs for the corresponding facts for the inverse Fourier transform.

2–3. For the Fourier transform of a product of f with t^n, we have

$$\mathcal{F}[t^n f(t)](\lambda) = \frac{1}{\sqrt{2\pi}} \int_{-\infty}^{\infty} t^n f(t)e^{-i\lambda t}\, dt.$$

Using the equation

$$t^n f(t)e^{-i\lambda t} = (i)^n \frac{d^n}{d\lambda^n}\{f(t)e^{-i\lambda t}\},$$

we obtain

$$\mathcal{F}[t^n f(t)](\lambda) = (i)^n \frac{d^n}{d\lambda^n}\left\{\frac{1}{\sqrt{2\pi}} \int_{-\infty}^{\infty} f(t)e^{-i\lambda t}\, dt\right\}$$

$$= (i)^n \frac{d^n}{d\lambda^n}\{\mathcal{F}[f](\lambda)\}.$$

The corresponding property for the inverse Fourier transform is proved similarly.

4–5. For the Fourier transform of the nth derivative of f, we have

$$\mathcal{F}[f^{(n)}(t)](\lambda) = \frac{1}{\sqrt{2\pi}} \int_{-\infty}^{\infty} f^{(n)}(t)e^{-i\lambda t}\, dt.$$

We now integrate by parts; that is,

$$\int_{-\infty}^{\infty} u\, dv = uv\Big|_{-\infty}^{\infty} - \int_{-\infty}^{\infty} v\, du$$

with $dv = f^{(n)}$ and $u = e^{-i\lambda t}$. As we will see, this process has the effect of transferring the derivatives from $f^{(n)}$ to $e^{-i\lambda t}$. Since f vanishes at $+\infty$

and $-\infty$ by hypothesis, there are no boundary terms. So we have

$$\frac{1}{\sqrt{2\pi}} \int_{-\infty}^{\infty} f^{(n)}(t)e^{-i\lambda t}\, dt = -\frac{1}{\sqrt{2\pi}} \int_{-\infty}^{\infty} f^{(n-1)}(t)\frac{d}{dt}\{e^{-i\lambda t}\}\, dt \quad \text{(by parts)}$$

$$= (i\lambda)\frac{1}{\sqrt{2\pi}} \int_{-\infty}^{\infty} f^{(n-1)}(t)e^{-i\lambda t}\, dt.$$

Note that integration by parts has reduced the number of derivatives on f by one [$f^{(n)}$ becomes $f^{(n-1)}$]. We have also gained one factor of $i\lambda$. By repeating this process $n-1$ additional times, we obtain

$$\frac{1}{\sqrt{2\pi}} \int_{-\infty}^{\infty} f^{(n)}(t)e^{-i\lambda t}\, dt = (i\lambda)^n\frac{1}{\sqrt{2\pi}} \int_{-\infty}^{\infty} f(t)e^{-i\lambda t}\, dt$$

$$= (i\lambda)^n \mathcal{F}[f](\lambda)$$

as desired. The proof for the corresponding fact for the inverse Fourier transform is similar.

6–7. The translation and rescaling properties can be combined into the following statement:

$$\mathcal{F}[f(bt-a)](\lambda) = \frac{1}{b}e^{-i\lambda a/b}\mathcal{F}[f](\lambda/b). \tag{2.6}$$

This equation can be established by the following change-of-variables argument:

$$\mathcal{F}[f(bt-a)](\lambda) = \frac{1}{\sqrt{2\pi}} \int_{-\infty}^{\infty} f(bt-a)e^{-i\lambda t}\, dt$$

$$= \frac{1}{\sqrt{2\pi}} \int_{-\infty}^{\infty} f(s)e^{-i\lambda(\frac{s+a}{b})}\frac{ds}{b}$$

where the second equality follows from the change of variables $s = bt - a$ [and so $t = (s+a)/b$ and $dt = \frac{ds}{b}$]. Rewriting the exponential on the right as

$$e^{-i\lambda(\frac{s+a}{b})} = e^{\frac{-i\lambda a}{b}}e^{-i\frac{\lambda}{b}s},$$

we obtain

$$\mathcal{F}[f(bt-a)](\lambda) = e^{\frac{-i\lambda a}{b}}\frac{1}{\sqrt{2\pi}} \int_{-\infty}^{\infty} f(s)e^{-i\frac{\lambda}{b}s}\frac{ds}{b}$$

$$= e^{\frac{-i\lambda a}{b}}\frac{1}{b}\mathcal{F}[f](\frac{\lambda}{b}).$$

8. The final part of the theorem, regarding the relationship of the Fourier and Laplace transforms, follows from their definitions and is left to the exercises. This completes the proof.   ♦

EXAMPLE 2.7

We illustrate the fourth property of Theorem 2.6 with the function in Example 2.4:

$$f(t) = \begin{cases} \sin 3t & \text{if } -\pi \leq t \leq \pi \\ 0 & \text{otherwise} \end{cases}$$

whose Fourier transform is

$$\widehat{f}(\lambda) = \frac{-3\sqrt{2}i\sin(\lambda\pi)}{\sqrt{\pi}(9 - \lambda^2)}.$$

The derivative of f is $3\cos 3t$ on $-\pi \leq t \leq \pi$, which is just a multiple of 3 times the function given in Example 2.3. Therefore, from Example 2.3,

$$\widehat{f'}(\lambda) = \frac{3\sqrt{2}\lambda\sin(\lambda\pi)}{\sqrt{\pi}(9 - \lambda^2)}. \tag{2.7}$$

The fourth property of Theorem 2.6 (with $n = 1$) states

$$\widehat{f'}(\lambda) = i\lambda\widehat{f}(\lambda) = -i\lambda\frac{3\sqrt{2}i\sin(\lambda\pi)}{\sqrt{\pi}(9 - \lambda^2)}.$$

This expression for $\widehat{f'}(\lambda)$ agrees with (2.7). ∎

EXAMPLE 2.8

In this example, we illustrate the scaling property:

$$\mathcal{F}[f(bt)](\lambda) = \frac{1}{b}\mathcal{F}[f](\frac{\lambda}{b}). \tag{2.8}$$

If $b > 1$, then the graph of $f(bt)$ is a compressed version of the graph of f. The dominant frequencies of $f(bt)$ are larger than those of f by a factor of b. This behavior is illustrated nicely by the function in Example 2.4:

$$f(t) = \begin{cases} \sin 3t & \text{if } -\pi \leq t \leq \pi \\ 0 & \text{otherwise} \end{cases}$$

whose graph is given in Figure 7. The graph of $f(2t)$ is given in Figure 9. Note that the frequency $f(2t)$ is double that of f.

Increasing the frequency of a signal has the effect of stretching the graph of its Fourier transform. In the preceding example, the dominant frequency of $f(t)$ is 3 whereas the dominant frequency of $f(2t)$ is 6. Thus the maximum value of $|\widehat{f}(\lambda)|$ occurs at $\lambda = 3$ (see Figure 8) whereas the maximum value of the Fourier transform of $f(2t)$ occurs at $\lambda = 6$ (see Figure 10). Thus the latter graph is obtained by stretching the former graph by a factor of 2. Note also that the graph of $\widehat{f}(\lambda/2)$ is obtained by stretching the graph of $\widehat{f}(\lambda)$ by a factor of 2.

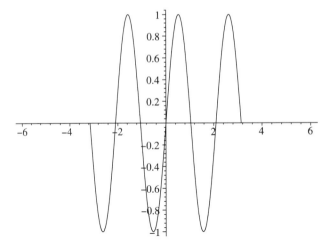

Figure 7 Plot of $\sin(3t)$, $-\pi \leq t \leq \pi$

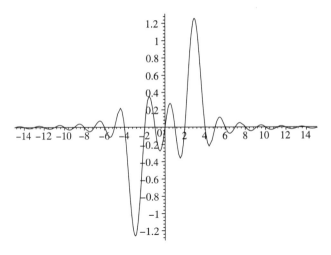

Figure 8 Plot of Fourier transform of $\sin(3t)$

This discussion illustrates the following geometrical interpretation of equation (2.8) in the case where $b > 1$: Compressing the graph of f speeds up the frequency and therefore stretches the graph of $\widehat{f}(\lambda)$. If $0 < b < 1$, then the graph of $f(bt)$ is stretched, which slows the frequency and therefore compresses the graph of $\widehat{f}(\lambda)$.   ∎

2.2.2 Fourier Transform of a Convolution

Now we examine how the Fourier transform behaves under convolutions. First, we give the definition of the convolution of two functions.

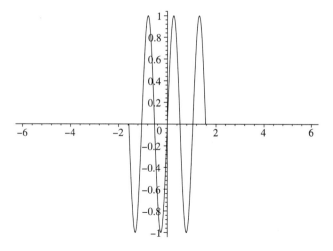

Figure 9 Plot of $\sin(6t)$ $-\pi/2 \le t \le \pi/2$

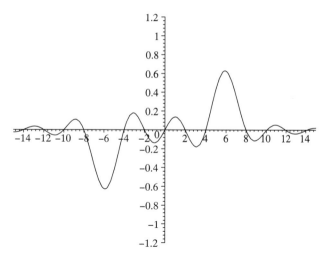

Figure 10 Plot of Fourier transform of $\sin(6t)$

DEFINITION 2.9 *Suppose f and g are two square-integrable functions. The convolution of f and g, denoted f * g, is defined by*

$$(f * g)(t) = \int_{-\infty}^{\infty} f(t - x)g(x)\,dx.$$

The preceding definition is equivalent to

$$(f * g)(t) = \int_{-\infty}^{\infty} f(x)g(t - x)\,dx$$

(perform the change of variables $y = t - x$ and then relabel the variable y back to x).

We have the following theorem on the Fourier transform of a convolution of two functions.

THEOREM 2.10 *Suppose f and g are two integrable functions. Then*

$$\mathcal{F}[f * g] = \sqrt{2\pi}\,\widehat{f} \cdot \widehat{g} \tag{2.9}$$

$$\mathcal{F}^{-1}[\widehat{f} \cdot \widehat{g}] = \frac{1}{\sqrt{2\pi}} f * g. \tag{2.10}$$

Proof To derive the first equation, we use the definitions of the Fourier transform and convolution:

$$\mathcal{F}[f * g](\lambda) = \frac{1}{\sqrt{2\pi}} \int_{-\infty}^{\infty} (f * g)(t) e^{-i\lambda t}\, dt$$

$$= \frac{1}{\sqrt{2\pi}} \int_{-\infty}^{\infty} \int_{-\infty}^{\infty} f(t - x) g(x)\, dx\, e^{-i\lambda t}\, dt.$$

We write $e^{-i\lambda t} = e^{-i\lambda(t-x)} e^{-i\lambda x}$. After switching the order of integration, we obtain

$$\mathcal{F}[f * g](\lambda) = \frac{1}{\sqrt{2\pi}} \int_{-\infty}^{\infty} \int_{-\infty}^{\infty} f(t - x) e^{-i\lambda(t-x)} g(x)\, dt\, e^{-i\lambda x}\, dx.$$

Letting $s = t - x$, we obtain

$$\mathcal{F}[f * g](\lambda) = \frac{1}{\sqrt{2\pi}} \int_{-\infty}^{\infty} \int_{-\infty}^{\infty} f(s) e^{-i\lambda s}\, ds\, g(x) e^{-i\lambda x}\, dx.$$

The right side can be rewritten as

$$\sqrt{2\pi} \left(\frac{1}{\sqrt{2\pi}} \int_{-\infty}^{\infty} f(s) e^{-i\lambda s}\, ds \right) \left(\frac{1}{\sqrt{2\pi}} \int_{-\infty}^{\infty} g(x) e^{-i\lambda x}\, dx \right),$$

which is $\sqrt{2\pi}\,\widehat{f}(\lambda)\widehat{g}(\lambda)$, as desired.

Equation (2.10) follows from equation (2.9) and the inverse formula for the Fourier transform as follows:

$$\sqrt{2\pi}\,\mathcal{F}^{-1}[\widehat{f}\widehat{g}] = \mathcal{F}^{-1}[\mathcal{F}(f * g)] \quad \text{from (2.9)}$$

$$= f * g \quad \text{from Theorem 2.1.}$$

This completes the proof of the theorem.   ◆

2.2.3 Adjoint of the Fourier Transform

Recall from Section 0.6.2 that the adjoint of a linear operator $T : V \mapsto W$ between two inner product spaces is an operator $T^* : W \mapsto V$ such that

$$\langle v, T^*(w) \rangle_V = \langle T(v), w \rangle_W.$$

In the next theorem, we show that the adjoint of the Fourier transform is the inverse of the Fourier transform.

THEOREM 2.11 *Suppose f and g are square integrable. Then*

$$\langle \mathcal{F}[f], g \rangle_{L^2} = \langle f, \mathcal{F}^{-1}[g] \rangle_{L^2}.$$

Proof We have

$$\langle \mathcal{F}[f], g \rangle_{L^2} = \int_{-\infty}^{\infty} \widehat{f}(\lambda) \overline{g(\lambda)} \, d\lambda$$

$$= \frac{1}{\sqrt{2\pi}} \int_{-\infty}^{\infty} \int_{-\infty}^{\infty} f(t) e^{-i\lambda t} \, dt \, \overline{g(\lambda)} \, d\lambda \quad \text{by definition of } \widehat{f}$$

$$= \int_{-\infty}^{\infty} f(t) \left(\frac{1}{\sqrt{2\pi}} \int_{-\infty}^{\infty} \overline{g(\lambda) e^{i\lambda t}} \, d\lambda \right) dt$$

(by switching the order of integration). The second integral (involving g) is $\overline{\mathcal{F}^{-1}[g](t)}$; therefore,

$$\langle \mathcal{F}[f], g \rangle_{L^2} = \int_{-\infty}^{\infty} f(t) \overline{\mathcal{F}^{-1}[g](t)} \, dt = \langle f, \mathcal{F}^{-1}[g] \rangle_{L^2}$$

as desired. ◆

2.2.4 Plancherel Formula

The Plancherel formula states that the Fourier transform preserves the L^2 inner product.

THEOREM 2.12 *Suppose f and g are square integrable. Then*

$$\langle \mathcal{F}[f], \mathcal{F}[g] \rangle_{L^2} = \langle f, g \rangle_{L^2} \tag{2.11}$$

$$\langle \mathcal{F}^{-1}[f], \mathcal{F}^{-1}[g] \rangle_{L^2} = \langle f, g \rangle_{L^2}. \tag{2.12}$$

In particular,

$$\|\mathcal{F}[f]\|_{L^2} = \|f\|_{L^2}. \tag{2.13}$$

Proof Equation (2.11) follows from Theorems 2.11 and 2.1 as follows:

$$\langle \mathcal{F}[f], \mathcal{F}[g] \rangle_{L^2} = \langle f, \mathcal{F}^{-1}\mathcal{F}[g] \rangle_{L^2} \qquad \text{(Theorem 2.11)}$$
$$= \langle f, g \rangle_{L^2} \qquad \text{(Theorem 2.1)}$$

as desired. Equation (2.12) can be established in a similar manner. Equation (2.13) follows from (2.11) with $f = g$.   ♦

Remark. The equation $\|\mathcal{F}(f)\| = \|f\|$ is analogous to equation (1.38) in Theorem 1.40 and is also referred to as Parseval's equation. This equation has the following interpretations. For a function, f, defined on $[-\pi, \pi]$, let $\mathcal{F}(f)(n)$ be its nth Fourier coefficient (except with the factor $1/\sqrt{2\pi}$ instead of $1/2\pi$):

$$\mathcal{F}(f)(n) = \frac{1}{\sqrt{2\pi}} \int_{-\pi}^{\pi} f(t)e^{-int}\, dt.$$

Equation (1.38) can be restated as

$$\|\mathcal{F}(f)\|_{l^2}^2 = \|f\|_{L^2[-\pi,\pi]}^2$$

which is analogous to (2.13) for the Fourier transform with l^2 and $L^2[-\pi, \pi]$ replaced by the L^2 norm on the entire real line. As in the case with Fourier series, Plancherel's formula states that the energy of a signal in the time domain, $\|f\|_{L^2}^2$, is the same as the energy in the frequency domain $\|\hat{f}\|_{L^2}^2$.

2.3   Linear Filters

2.3.1   Time Invariant Filters

The Fourier transform plays a central role in the design of filters. A filter can be thought of as a "black box" that takes an input signal, processes it, and then returns an output signal that in some way modifies the input. One example of a filter is a device that removes noise from a signal.

From a mathematical point of view, a signal is a function $f : R \mapsto C$ that is usually piecewise continuous. A filter is a transformation L that maps a signal, f, into another signal $\tilde{f}$. This transformation must satisfy the following two properties in order to be a *linear filter* :

- Additivity: $L[f + g] = L[f] + L[g]$

- Homogeneity: $L[cf] = cL[f]$, where c is a constant

There is another property common to most filters. If we play an old, scratchy record starting at 3 P.M. today and put the signal through a noise-reducing filter, we will hear the cleaned-up output, at roughly the same time as we play the record. If we play the same record at 10 A.M. tomorrow morning, and use the same filter, we should hear the identical output, again at roughly the same time. This property is called *time invariance*. To formulate this concept, we introduce the following notation: For a function $f(t)$ and a real number a, let $f_a(t) = f(t - a)$. Thus f_a is a time shift, by a units, of the signal f.

DEFINITION 2.13 *A transformation L (mapping signals to signals) is said to be* time invariant *if for any signal f and any real number a, $L[f_a](t) = (Lf)(t-a)$ for all t (or $L[f_a] = (Lf)_a$). In words, L is time invariant if the time-shifted input signal $f(t - a)$ is transformed by L into the time-shifted output signal $(Lf)(t - a)$. (See Figure 11.)*

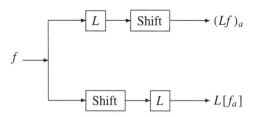

Figure 11 L is time invariant if the upper and lower outputs are the same

---

EXAMPLE 2.14

Let $l(t)$ be a function that has finite support [i.e., $l(t)$ is zero outside of a finite t-interval]. For a signal f, let

$$(Lf)(t) = (l * f)(t) = \int_{-\infty}^{\infty} l(t - x)f(x)\, dx \quad \text{for each } t.$$

This linear operator is time invariant because for any $a \in R$

$$(Lf)(t - a) = \int_{-\infty}^{\infty} l(t - a - x)f(x)\, dx$$

$$= \int_{-\infty}^{\infty} l(t - y)f(y - a)\, dy \quad \text{(by letting } y = a + x)$$

$$= \int_{-\infty}^{\infty} l(t - y)f_a(y)\, dy$$

$$= L[f_a](t).$$

Thus, $(Lf)(t - a) = L[f_a](t)$ and so L is time invariant. ∎

Not every linear transformation has this property, as the following example shows.

EXAMPLE 2.15

Let

$$(Lf)(t) = \int_0^t f(\tau)\,d\tau.$$

On one hand, we have

$$L[f_a](t)) = \int_0^t f(\tau - a)\,d\tau$$
$$= \int_{-a}^{t-a} f(\hat{\tau})\,d\hat{\tau} \quad \text{by letting } \hat{\tau} = \tau - a.$$

On the other hand,

$$(Lf)(t - a) = \int_0^{t-a} f(\tau)\,d\tau.$$

Since $L[f_a](t)$ and $(Lf)(t - a)$ are not the same (for $a \neq 0$), L is *not* time invariant. ■

The next lemma and theorem show that the convolution in Example 2.14 is typical of time invariant linear filters. We start by computing $L(e^{i\lambda t})$.

LEMMA 2.16 *Let L be a linear, time-invariant transformation and let λ be any fixed real number. Then there is a function h with*

$$L(e^{i\lambda t}) = \sqrt{2\pi}\,\hat{h}(\lambda)e^{i\lambda t} \qquad \text{(t is the variable).}$$

Remark. Note that the input signal $e^{i\lambda t}$ is a (complex-valued) sinusoidal signal with frequency λ. This lemma states that the output signal from a time invariant filter of a sinusoidal input signal is also sinusoidal with the same frequency.

Proof Our proof is somewhat informal in order to explain clearly the essential idea. Let $h^\lambda(t) = L(e^{i\lambda t})$. Since L is time invariant, we have

$$L[e^{i\lambda(t-a)}] = h^\lambda(t - a) \tag{2.14}$$

for each real number a. Since L is linear, we also have

$$L[e^{i\lambda(t-a)}] = L[e^{-i\lambda a}e^{i\lambda t}]$$
$$= e^{-i\lambda a}L[e^{i\lambda t}] \quad (L \text{ is linear}).$$

Thus

$$L[e^{i\lambda(t-a)}] = e^{-i\lambda a}h^\lambda(t). \tag{2.15}$$

Comparing (2.14) and (2.15), we find

$$h^\lambda(t - a) = e^{-i\lambda a} h^\lambda(t).$$

Since a is arbitrary, we may set $a = t$, yielding $h^\lambda(0) = e^{-i\lambda t} h^\lambda(t)$; solving for $h^\lambda(t)$, we obtain

$$L[e^{i\lambda t}] = h^\lambda(t) = h^\lambda(0)e^{i\lambda t}.$$

Letting $\hat{h}(\lambda) = h^\lambda(0)/\sqrt{2\pi}$ completes the proof.   ♦

The function $\hat{h}(\lambda)$ determines L. To see this, we first use the Fourier inversion theorem (Theorem 2.1):

$$f(t) = \mathcal{F}^{-1}[\hat{f}] = \frac{1}{\sqrt{2\pi}} \int_{-\infty}^{\infty} \hat{f}(\lambda)e^{i\lambda t} d\lambda.$$

Then we apply L to both sides:

$$(Lf)(t) = L\left[\frac{1}{\sqrt{2\pi}} \int_{-\infty}^{\infty} \hat{f}(\lambda)e^{i\lambda t} d\lambda\right]. \tag{2.16}$$

The integral on the right can be approximated by a Riemann sum:

$$L\left[\frac{1}{\sqrt{2\pi}} \int_{-\infty}^{\infty} \hat{f}(\lambda)e^{i\lambda t} d\lambda\right] \approx L\left[\frac{1}{\sqrt{2\pi}} \sum_j \hat{f}(\lambda_j)e^{i\lambda_j t}\Delta\lambda\right]. \tag{2.17}$$

Since L is linear, we can distribute L across the sum:

$$L\left[\frac{1}{\sqrt{2\pi}} \sum_j \hat{f}(\lambda_j)e^{i\lambda_j t}\Delta\lambda\right] = \frac{1}{\sqrt{2\pi}} \sum_j \hat{f}(\lambda_j)L\left[e^{i\lambda_j t}\right]\Delta\lambda. \tag{2.18}$$

As the partition gets finer, the Riemann sum on the right becomes an integral, and so from (2.16), (2.17), and (2.18), we obtain

$$(Lf)(t) = \frac{1}{\sqrt{2\pi}} \int_{-\infty}^{\infty} \hat{f}(\lambda)L[e^{i\lambda t}](t)\, d\lambda$$

$$= \frac{1}{\sqrt{2\pi}} \int_{-\infty}^{\infty} \hat{f}(\lambda)\left(\sqrt{2\pi}\hat{h}(\lambda)\right)e^{i\lambda t}\, d\lambda \quad \text{by Lemma 2.16}$$

$$= \sqrt{2\pi}\mathcal{F}^{-1}[\hat{f}(\lambda)\hat{h}(\lambda)](t) \quad \text{by definition of inverse Fourier transform}$$

$$= (f * h)(t) \quad \text{by Theorem 2.10.}$$

Even though the preceding argument is not totally rigorous, the result is true with very few restrictions on either L or the space of signals being considered (see the text on Fourier analysis by Stein and Weiss [19] for more details). We summarize this discussion in the following theorem.

Theorem 2.17 *Let L be a linear, time invariant transformation on the space of signals that are piecewise continuous functions. Then there exists an integrable function, h, such that*

$$L(f) = f * h$$

for all signals f.

Physical Interpretation. Both $h(t)$ and $\hat{h}(\lambda)$ have physical interpretations. Assume that $h(t)$ is continuous and that δ is a small positive number. We apply L to the following impulse signal:

$$f_\delta(t) = \begin{cases} 1/(2\delta) & \text{if } -\delta \le t \le \delta \\ 0 & \text{otherwise} \end{cases}$$

whose graph is given in Figure 12. If $\delta > 0$ is small, then f_δ represents a signal that is very strong but only lasts a short period of time (such as the sound signal generated by a hammer blow). Note that $\int_{-\delta}^{\delta} f_\delta(t)\, dt = 1$. Applying L to f_δ, we obtain

$$(Lf_\delta)(t) = f_\delta * h(t)$$

$$= \int_{-\infty}^{\infty} f_\delta(\tau) h(t - \tau)\, d\tau$$

$$= \int_{-\delta}^{\delta} f_\delta(\tau) h(t - \tau)\, d\tau \quad (\text{since } f_\delta(\tau) = 0 \text{ for } |\tau| \ge \delta).$$

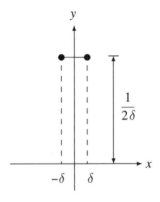

Figure 12 Graph of f_δ

Since h is continuous, $h(t - \tau)$ is approximately equal to $h(t)$ for $|\tau| \leq \delta$. Therefore,

$$(Lf_\delta)(t) \approx h(t) \underbrace{\int_{-\delta}^{\delta} f_\delta(\tau) \, d\tau}_{1} = h(t).$$

Thus $h(t)$ is the approximate response from applying L to an input signal that is an impulse. For that reason $h(t)$ is called the *impulse response function*.

We have already seen that $L[e^{i\lambda t}] = \sqrt{2\pi}\hat{h}(\lambda)e^{i\lambda t}$. Thus up to a constant factor, $\hat{h}(\lambda)$ is the amplitude of the response to a "pure frequency" signal $e^{i\lambda t}$; $\hat{h}$ is called the *system function*.

2.3.2 Causality and the Design of Filters

Designing a time invariant filter is equivalent to constructing the impulse function, h, since any such filter can be written as

$$Lf = f * h$$

by Theorem 2.17. The construction of h depends on what the filter is designed to do. In this section, we consider filters that reduce high frequencies. Such filters are called *low-pass* filters.

Taking the Fourier transform of both sides of $Lf = f * h$ and using Theorem 2.10 yields

$$\widehat{L[f]}(\lambda) = \sqrt{2\pi}\hat{f}(\lambda)\hat{h}(\lambda).$$

A Faulty Filter. Suppose we wish to remove all frequency components from the signal f that lie beyond some cutoff frequency λ_c. As a natural first attempt, we choose an h whose Fourier transform is zero outside of the interval $-\lambda_c \leq \lambda \leq \lambda_c$:

$$\hat{h}_{\lambda_c}(\lambda) := \begin{cases} 1/\sqrt{2\pi} & \text{if } |\lambda| \leq \lambda_c \\ 0 & \text{if } |\lambda| > \lambda_c \end{cases} \tag{2.19}$$

(the choice of constant $1/\sqrt{2\pi}$ is for convenience in later calculations). Since $\widehat{Lf(\lambda)} = \sqrt{2\pi}\widehat{f(\lambda)}\widehat{h(\lambda)}$ is zero for $|\lambda| > \lambda_c$, this filter appears to remove the unwanted frequencies (above λ_c) from the signal f. However, we will see that this filter is flawed in other respects.

The impulse response function corresponding to the system function $\hat{h}_{\lambda_c}$ is easy to calculate:

$$h_{\lambda_c}(t) = \mathcal{F}^{-1}[\hat{h}_{\lambda_c}] \quad \text{(by Theorem 2.1)}$$

$$= (1/\sqrt{2\pi}) \int_{-\infty}^{\infty} \hat{h}_{\lambda_c}(\lambda)e^{i\lambda t}d\lambda$$

$$= \frac{1}{2\pi} \int_{-\lambda_c}^{\lambda_c} e^{i\lambda t}d\lambda \quad \text{from (2.19)}$$

$$= \left[\frac{e^{i\lambda t}}{2it\pi}\right]_{\lambda=-\lambda_c}^{\lambda=\lambda_c}$$

$$= \frac{e^{i\lambda_c t} - e^{-i\lambda_c t}}{2it\pi}.$$

Therefore,

$$h_{\lambda_c}(t) = \frac{\sin(\lambda_c t)}{\pi t}. \tag{2.20}$$

Now we filter the following simple input function:

$$f_{t_c}(t) := \begin{cases} 1 & \text{if } 0 \leq t \leq t_c \\ 0 & \text{if } t < 0 \text{ or } t > t_c. \end{cases}$$

Think of f_{t_c} as a signal that is on for t between 0 and t_c and off at other times. The effect of this filter on the signal f_{t_c} is

$$(Lf_{t_c})(t) = (f_{t_c} * h_{\lambda_c})(t)$$

$$= \int_{-\infty}^{\infty} f_{t_c}(\tau)h_{\lambda_c}(t - \tau)\,d\tau$$

$$= \int_{0}^{t_c} \frac{\sin(\lambda_c(t - \tau))}{\pi(t - \tau)}\,d\tau \quad \text{[from (2.20)]}$$

$$= \frac{1}{\pi} \int_{\lambda_c(t-t_c)}^{\lambda_c t} \frac{\sin u}{u}\,du \quad \text{[with } u = \lambda_c(t - \tau)\text{]}$$

$$= \frac{1}{\pi} \{\text{Si}(\lambda_c t) - \text{Si}(\lambda_c(t - t_c))\}$$

where $\text{Si}(z) = \int_0^z \frac{\sin u}{u} du$. A plot of $(L_{\lambda_c} f_{t_c})(t) = \frac{1}{\pi}\{\text{Si}(\lambda_c t) - \text{Si}(\lambda_c(t - t_c))\}$ with t_c and λ_c both 1 is given in Figure 13. Note that the graph of the output signal is nonzero for $t < 0$ whereas the input signal, $f_{t_c}(t)$, is zero for $t < 0$. This indicates that the output signal occurs before the input signal has arrived!

Clearly, a filter cannot be physically constructed to produce an output signal before receiving an input signal. Thus, our first attempt at constructing a filter by using the function h_{λ_c} is not practical. The following definition and theorem characterize a more realistic class of filters.

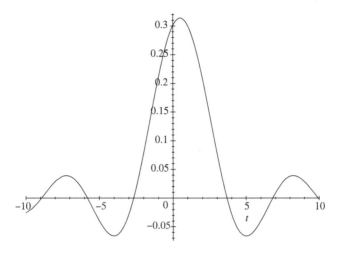

Figure 13 Graph of $\frac{1}{\pi}\left\{\text{Si}(t) - \text{Si}(t-1)\right\}$

Causal Filters

DEFINITION 2.18 *A causal filter is one for which the output signal begins* after *the input signal has started to arrive.*

The following result tells us which filters are causal.

THEOREM 2.19 *Let L be a time invariant filter with response function h (i.e., $Lf = f * h$). L is a causal filter if and only if $h(t) = 0$ for all $t < 0$.*

Proof We prove that if $h(t) = 0$ for all $t < 0$, then the corresponding filter is causal. We leave the converse as an exercise (see Exercise 8). We first show that if $f(t) = 0$ for $t < 0$, then $(Lf)(t) = 0$ for $t < 0$. We have

$$(Lf)(t) = (f * h)(t) = \int_0^\infty f(\tau)h(t - \tau)\, d\tau,$$

where the lower limit in the integral is 0 because $f(\tau) = 0$ when $\tau < 0$. If $t < 0$ and $\tau \geq 0$, then $t - \tau < 0$ and so $h(t - \tau) = 0$, by hypothesis. Therefore, $(Lf)(t) = 0$ for $t < 0$. We have therefore shown that if $f(t) = 0$ for $t < 0$, then $(Lf)(t) = 0$ for $t < 0$. In other words, if the input signal does not arrive until $t = 0$, the output of the filter also does not begin until $t = 0$.

Suppose, more generally, that the input signal f does not arrive until $t = a$. To show that L is causal, we must show that Lf does not begin until $t = a$. Let

$g(t) = f(t + a)$. Note that the signal $g(t)$ begins at $t = 0$. From the previous paragraph, $(Lg)(t)$ does not begin until $t = 0$. Since $f(t) = g(t - a) = g_a(t)$, we have

$$(Lf)(t) = (Lg_a)(t)$$
$$= (Lg)(t - a) \quad \text{by the time invariance of } L.$$

Since $(Lg)(\tau)$ does not begin until $\tau = 0$, we see that $(Lg)(t - a)$ does not begin until $t = a$. Thus $(Lf)(t) = (Lg)(t - a)$ does not begin until $t = a$, as desired. ♦

Theorem 2.19 applies to the response function, but it also gives us important information about the system function, $\hat{h}(\lambda)$. By the definition of the Fourier transform,

$$\hat{h}(\lambda) = \frac{1}{\sqrt{2\pi}} \int_{-\infty}^{\infty} h(t)e^{-i\lambda t} dt.$$

If the filter associated to h is causal, Theorem 2.19 implies $h(t) = 0$ for $t < 0$ and so

$$\hat{h}(\lambda) = \frac{1}{\sqrt{2\pi}} \int_{0}^{\infty} h(t)e^{-i\lambda t} dt$$
$$= \mathcal{L}[h(t)/\sqrt{2\pi}](i\lambda) \quad \text{(where } \mathcal{L} = \text{Laplace transform).}$$

We summarize this discussion in the following theorem.

THEOREM 2.20 *Suppose L is a causal filter with response function h. Then the system function associated with L is*

$$\hat{h}(\lambda) = \frac{\mathcal{L}[h](i\lambda)}{\sqrt{2\pi}}$$

where $\mathcal{L}$ is the Laplace transform.   ♦

EXAMPLE 2.21

One of the older causal, noise-reducing filters is the *Butterworth* filter [14]. It is constructed using the previous theorem with

$$h(t) = \begin{cases} Ae^{-\alpha t} & \text{if } t \geq 0 \\ 0 & \text{if } t < 0 \end{cases}$$

where A and α are parameters. Its Fourier transform is given by

$$\hat{h}(\lambda) = \frac{1}{\sqrt{2\pi}} (\mathcal{L}h)(i\lambda) = \frac{A}{\sqrt{2\pi}(\alpha + i\lambda)}$$

(see Exercise 10). Note that $\hat{h}(\lambda)$ decays as $\lambda \mapsto \infty$, thus diminishing the high-frequency components of the filtered signal $(\widehat{Lf})(\lambda) = \hat{h}(\lambda)\hat{f}(\lambda)$.

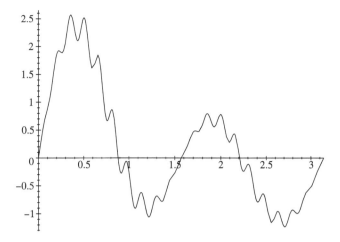

Figure 14 Graph of $e^{-t/3}\left(\sin 2t + 2\sin 4t + 0.4\sin 2t\sin 40t\right)$

Consider the signal given by

$$f(t) = e^{-t/3}\left(\sin 2t + 2\sin 4t + 0.4\sin 2t\sin 40t\right) \quad 0 \le t \le \pi$$

whose graph is given in Figure 14. We wish to filter the noise that vibrates with frequency approximately 40. At the same time, we do not want to disturb the basic shape of this signal, which vibrates in the frequency range of 2 to 4. By choosing $A = \alpha = 10$, $\widehat{h}(\lambda)$ is close to $\widehat{h}(0) = 1/\sqrt{2\pi}$ for $|\lambda| \le 4$; but $|\widehat{h}(\lambda)|$ is small (less than 0.1) when $\lambda \ge 40$. Thus filtering by h with this choice of parameters A and α should preserve the low frequencies (frequency ≤ 4) while damping the high frequencies (≥ 40). A plot of the filtered signal $(f * h)(t)$ for $0 \le t \le \pi$ is given in Figure 15. Most of the high-frequency noise has been filtered. Most of the low-frequency components of this signal have been preserved. ∎

2.4 The Sampling Theorem

In this section, we examine a class of signals (i.e., functions) whose Fourier transform is zero outside a finite interval $[-\Omega, \Omega]$; these are (frequency) *band-limited functions*. For instance, the human ear can only hear sounds with frequencies less than 20 kHz (1 kHz = 1000 cycles per second). Thus, even though we *make* sounds with higher frequencies, anything above 20 kHz can't be heard. Telephone conversations are thus effectively band-limited signals. We will show below that a band-limited signal can be reconstructed from its values (or *samples*) at regularly spaced times. This result is basic in continuous-to-digital signal processing.

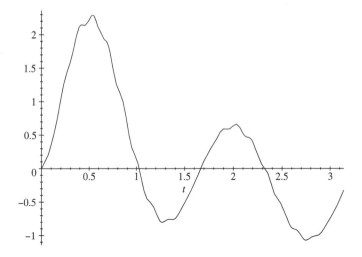

Figure 15 Graph of the filtered signal from Figure 14

DEFINITION 2.22 *A function f is said to be frequency band-limited if there exists a constant $\Omega > 0$ such that*

$$\hat{f}(\lambda) = 0 \quad \text{for } |\lambda| > \Omega.$$

When Ω is the smallest frequency for which the preceding equation is true, the natural frequency $\nu := \frac{\Omega}{2\pi}$ is called the Nyquist frequency, *and $2\nu = \frac{\Omega}{\pi}$ is the* Nyquist rate.

THEOREM 2.23 (Shannon-Whittaker Sampling Theorem) *Suppose that $\hat{f}(\lambda)$ is piecewise smooth, continuous, and that $\hat{f}(\lambda) = 0$ for $|\lambda| > \Omega$, where Ω is some fixed, positive frequency. Then $f = \mathcal{F}^{-1}[\hat{f}]$ is completely determined by its values at the points $t_j = j\pi/\Omega$, $j = 0, \pm 1, \pm 2, \ldots$. More precisely, f has the following series expansion:*

$$f(t) = \sum_{j=-\infty}^{\infty} f(j\pi/\Omega) \frac{\sin(\Omega t - j\pi)}{\Omega t - j\pi} \tag{2.21}$$

where the series on the right converges uniformly.

This is a remarkable theorem! Let's look at how we might use it to transmit *several* phone conversations simultaneously on a single wire (or channel). The maximum possible (natural) frequency that we can hear is about 20 kHz, so a phone conversation is effectively a band-limited signal. In fact, the dominant frequencies in most phone conversations are below 1 kHz, which we will take as our Nyquist frequency. The Nyquist *rate* $\nu = \frac{\Omega}{\pi}$ is then double this, or 2 kHz; so we need to sample the signal every $\frac{1}{2}$ millisecond. How many phone conversations can we send in this manner? Transmission lines typically send about 56 thousand bits of information per second. If each sample of information can be represented by 7 bits, then we can transmit 8 thousand samples per second, or 8 every millisecond, or 4 every half-millisecond. By tagging and interlacing signals, we can transmit the samples from four conversations. At the receiving end, we can use the series in equation (2.21) to reconstruct the signal, with the samples being $f(\frac{1}{2}j)$ for j an integer and time in milliseconds (only a finite number of these occur during the life of a phone conversation— unless perhaps a teenager is on the line). Here is the proof of the theorem.

Proof   Using Theorem 1.20 (with $a = \Omega$ and $t = \lambda$), we expand $\hat{f}(\lambda)$ in a Fourier series on the interval $[-\Omega, \Omega]$:

$$\hat{f}(\lambda) = \sum_{k=-\infty}^{\infty} c_k e^{i\pi k\lambda/\Omega}, \quad c_k = \frac{1}{2\Omega} \int_{-\Omega}^{\Omega} \hat{f}(\lambda)e^{-i\pi k\lambda/\Omega}d\lambda.$$

Since $\hat{f}(\lambda) = 0$ for $|\lambda| \geq \Omega$, the limits in the integrals defining c_k can be changed to $-\infty \ldots \infty$:

$$c_k = \frac{\sqrt{2\pi}}{2\Omega} \frac{1}{\sqrt{2\pi}} \int_{-\infty}^{\infty} \hat{f}(\lambda)e^{-i\pi k\lambda/\Omega}d\lambda.$$

By Theorem 2.1,

$$c_k = \frac{\sqrt{2\pi}}{2\Omega} f(-k\pi/\Omega).$$

If we use this expression for c_k in the preceding series, and if at the same time we change the summation index from k to $j = -k$, we obtain

$$\hat{f}(\lambda) = \sum_{j=-\infty}^{\infty} \frac{\sqrt{2\pi}}{2\Omega} f(j\pi/\Omega)e^{-i\pi j\lambda/\Omega}. \tag{2.22}$$

Since $\hat{f}$ is a continuous, piecewise smooth function, the series (2.22) converges uniformly by Theorem 1.30. Using Theorem 2.1,

$$f(t) = \frac{1}{\sqrt{2\pi}} \int_{-\infty}^{\infty} \hat{f}(\lambda)e^{i\lambda t} \, d\lambda$$

$$= \frac{1}{\sqrt{2\pi}} \int_{-\Omega}^{\Omega} \hat{f}(\lambda)e^{i\lambda t} \, d\lambda \quad \text{since } \hat{f}(\lambda) = 0, |\lambda| \geq \Omega.$$

Using (2.22) for $\hat{f}$ and interchanging the order of integration and summation, we obtain

$$f(t) = \sum_{j=-\infty}^{\infty} \frac{\sqrt{2\pi}}{2\Omega} f(j\pi/\Omega)\frac{1}{\sqrt{2\pi}} \int_{-\Omega}^{\Omega} e^{-i\pi j\lambda/\Omega+i\lambda t}d\lambda. \tag{2.23}$$

The integral in (2.23) is

$$\int_{-\Omega}^{\Omega} e^{-i\pi j\lambda/\Omega + i\lambda t} d\lambda = 2\frac{\Omega \sin(t\Omega - j\pi)}{t\Omega - j\pi}.$$

After simplifying (2.23), we obtain (2.21), which completes the proof.   ◆

The convergence rate in (2.21) is rather slow since the coefficients (in absolute value) decay like $1/j$. The convergence rate can be increased so that the terms behave like $1/j^2$ or better, by a technique called *oversampling*, which is discussed in Exercise 14. At the opposite extreme, if a signal is sampled below the Nyquist rate, then the signal reconstructed via equation (2.21) will not only be missing high frequency components, but it will also have the energy in those components transferred to low frequencies that may not have been in the signal at all. This is a phenomenon called *aliasing*.

EXAMPLE 2.24

Consider the function f defined by

$$\widehat{f}(\lambda) := \begin{cases} \sqrt{2\pi}(1 - \lambda^2) & \text{if } |\lambda| \leq 1 \\ 0 & \text{if } |\lambda| > 1. \end{cases}$$

We can calculate $f(t) = \mathcal{F}^{-1}[\widehat{f}]$ by computing

$$f(t) = \frac{1}{\sqrt{2\pi}} \int_{-\infty}^{\infty} \widehat{f}(\lambda)e^{i\lambda t}\, d\lambda$$

$$= \int_{-1}^{1} (1 - \lambda^2)e^{i\lambda t}\, d\lambda$$

$$= \frac{4\sin(t) - 4\cos(t)\,t}{t^3}.$$

The last equality can be obtained by integration by parts (or by using your favorite computer algebra system). The plot of f is given in Figure 16.

Since $\widehat{f}(\lambda) = 0$ for $|\lambda| > 1$, the frequency Ω from the sampling theorem can be chosen to be *any* number that is greater than or equal to 1. With $\Omega = 1$, we graph the partial sum of the first 30 terms in the series given in the sampling theorem in Figure 17; note that the two graphs are nearly identical.   ■

2.5   The Uncertainty Principle

In this section, we present the uncertainty principle, which in words states that a function cannot simultaneously have restricted support in time as well as in frequency. To explain these ideas, we need a definition.

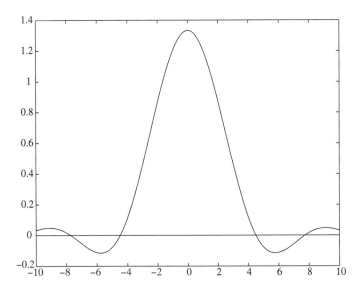

Figure 16 Graph of f for Example 2.24

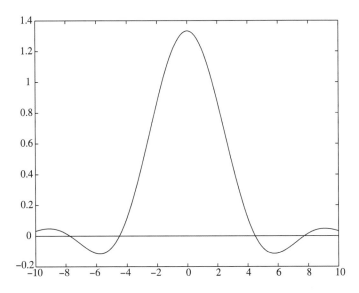

Figure 17 Graph of partial series in sampling theorem with $\Omega = 1$

DEFINITION 2.25 *Suppose f is a function in $L^2(R)$. The* dispersion *of f about the point $a \in R$ is the quantity*

$$\Delta_a f = \frac{\int_{-\infty}^{\infty} (t-a)^2 |f(t)|^2 \, dt}{\int_{-\infty}^{\infty} |f(t)|^2 \, dt}.$$

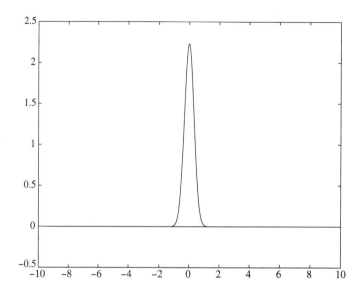

Figure 18 Small dispersion of f_5

The dispersion of f about a point a measures the deviation or spread of its graph from $t = a$. This dispersion will be small if the graph of f is concentrated near $t = a$, as in Figure 18 (with $a = 0$). The dispersion will be larger if the graph of f spreads out away from $t = a$ as in Figure 19 (with $a = 0$).

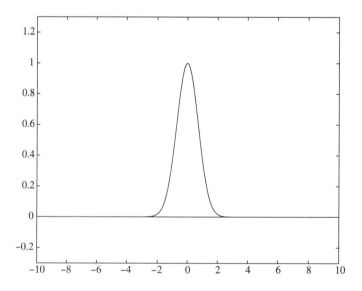

Figure 19 Larger dispersion of f_1

Another description of the dispersion is related to statistics. Think of the function $\frac{|f(t)|^2}{\int |f|^2}$ as a probability density function (this nonnegative function has integral equal to one, which is the primary requirement for a probability density function). If a is the mean of this density, then $\Delta_a f$ is just the variance.

Applying the preceding definition of dispersion to the Fourier transform of f gives

$$\Delta_\alpha \widehat{f} = \frac{\int_{-\infty}^{\infty} (\lambda - \alpha)^2 |\widehat{f}(\lambda)|^2 \, d\lambda}{\int_{-\infty}^{\infty} |\widehat{f}(\lambda)|^2 \, d\lambda}.$$

By the Plancherel formula (Theorem 2.12), the denominators in the dispersions of f and $\widehat{f}$ are the same.

If the dispersion of $\widehat{f}$ about $\lambda = \alpha$ is small, then the frequency range of f is concentrated near $\lambda = \alpha$.

Now we are ready to state the uncertainty principle.

THEOREM 2.26 (Uncertainty Principle) *Suppose f is a function in $L^2(R)$ that vanishes at $+\infty$ and $-\infty$. Then*

$$\Delta_a f \cdot \Delta_\alpha \widehat{f} \geq \frac{1}{4} \tag{2.24}$$

for all points $a \in R$ and $\alpha \in R$.

One consequence of the uncertainty principle is that the dispersion of f about any a (i.e., $\Delta_a f$) and the dispersion of the Fourier transform of f about any frequency α (i.e., $\Delta_\alpha \widehat{f}$) cannot simultaneously be small. The graph in Figure 18 offers an intuitive explanation. This graph is concentrated near $t = 0$, and therefore its dispersion about $t = 0$ is small. However, this function changes rapidly, and therefore it will have large-frequency components in its Fourier transform. Thus, the dispersion of $\widehat{f}$ about any frequency value will be large. This is illustrated by the wide spread of the graph of its Fourier transform, given in Figure 20.

For a more quantitative viewpoint, suppose

$$f_s(x) = \sqrt{s} e^{-sx^2}.$$

The graphs of f_s for $s = 5$ and $s = 1$ are given in Figures 18 and 19 respectively. Note that as s increases, the exponent becomes more negative and therefore the dispersion of f_s decreases (i.e., the graph of f_s becomes more concentrated near the origin).

The Fourier transform of f_s is

$$\widehat{f}_s(\lambda) = \frac{1}{\sqrt{2}} e^{\frac{-\lambda^2}{4s}}$$

(see Exercise 6). Except for the constant in front, the Fourier transform of f_s has the same general negative exponential form as does f_s. Thus the graph of $\widehat{f}_s$ has

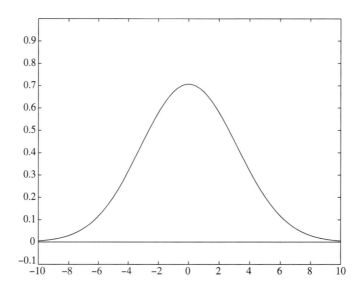

Figure 20 Large-frequency dispersion of $\hat{f}_5$

the same general shape as does the graph of f_s. There is one notable difference: The factor s appears in the denominator of the exponent in $\hat{f}_s$ instead of the numerator of the exponent (as is the case for f_s). Therefore, as s increases, the dispersion of $\hat{f}_s$ also increases (instead of decreasing as it does for f_s). In particular, it is not possible to choose a value of s that makes the dispersions of both f_s and $\hat{f}_s$ simultaneously small.

Proof of Uncertainty Principle   We first claim that the following identity holds:

$$\left\{ (\frac{d}{dt} - i\alpha)(t - a) \right\} f - \left\{ (t - a)(\frac{d}{dt} - i\alpha) \right\} f = f. \tag{2.25}$$

Here, α and a are real constants. Written out in detail, the left side is

$$\frac{d}{dt} \{(t - a)f\} - i\alpha(t - a)f - (t - a)(f' - i\alpha f).$$

After using the product rule for the first term and then simplifying, the result is f, which establishes (2.25).

Note that (2.25) remains valid after dividing both sides by the constant $\|f\| = \|f\|_{L^2}$. Since the L^2-norm of $f/\|f\|$ is 1, we may as well assume from the start that $\|f\| = 1$ (just by relabeling $f/\|f\|$ as f).

Now take the L^2 inner product of both sides of (2.25) with f. The result is

$$\left\langle (\frac{d}{dt} - i\alpha)\{(t - a)f(t)\}, f(t) \right\rangle - \left\langle (t - a)(\frac{d}{dt} - i\alpha)f(t), f(t) \right\rangle = \|f\|^2 = 1. \tag{2.26}$$

Both terms on the left involve integrals (from $-\infty$ to ∞). We use integration by parts on the first integral on the left and use the assumption that $f(-\infty) = f(\infty) = 0$. The result is

$$\left\langle (t-a)f(t), (-\frac{d}{dt} + i\alpha)f(t) \right\rangle - \left\langle (\frac{d}{dt} - i\alpha)f(t), (t-a)f(t) \right\rangle = 1 \quad (2.27)$$

(the details of which we ask you to carry out in Exercise 9).

From (2.27) and the triangle inequality, we obtain

$$1 \leq \left| \left\langle (t-a)f(t), (-\frac{d}{dt} + i\alpha)f(t) \right\rangle \right| + \left| \left\langle (\frac{d}{dt} - i\alpha)f(t), (t-a)f(t) \right\rangle \right|.$$

Now apply Schwarz's inequality (see Theorem 0.11) to the preceding two inner products:

$$1 \leq 2\|(\frac{d}{dt} - i\alpha)f(t)\| \, \|(t-a)f(t)\|.$$

Next, we apply the Plancherel formula (Theorem 2.12) and the fourth property listed in Theorem 2.6 to obtain

$$\|(\frac{d}{dt} - i\alpha)f(t)\| = \|(\lambda - \alpha)\hat{f}(\lambda)\|.$$

Combining this equation with the previous inequality, we get

$$\|(\lambda - \alpha)\hat{f}(\lambda)\| \, \|(t-a)f(t)\| \geq \frac{1}{2}.$$

Since $\|f\|_{L^2} = 1 = \|\hat{f}\|_{L^2}$,

$$\Delta_a(f) = \|(t-a)f\|_{L^2}^2 \quad \text{and} \quad \Delta_\alpha \hat{f} = \|(\lambda - \alpha)\hat{f}\|_{L^2}^2.$$

Therefore, squaring both sides of this inequality yields

$$\Delta_\alpha \hat{f} \, \Delta_a f \geq \frac{1}{4},$$

which completes the proof of the uncertainty principle.[1]   ◆

2.6   Exercises

1. Let

$$f(t) = \begin{cases} \cos(3t) & \text{for } -\pi \leq t \leq \pi \\ 0 & \text{otherwise.} \end{cases}$$

---

[1]This proof is based on that given for the quantum mechanical version by H. P. Robertson, *Physical Review* **34** (1929), 163–164.

Show that

$$\int_{-\pi}^{\pi} \cos(mt)\cos(\lambda t)\, dt = -2\frac{(-1)^m \lambda \sin(\pi\lambda)}{m^2 - \lambda^2}$$

where m is an integer and $\lambda \neq m$. *Hint*: Sum the identities

$$\cos(u+v) = \cos u \cos v - \sin u \sin v$$
$$\cos(u-v) = \cos u \cos v + \sin u \sin v.$$

Use this integral to show that

$$\widehat{f}(\lambda) = \frac{-\sqrt{2}\lambda \sin(\lambda\pi)}{\sqrt{\pi}(\lambda^2 - 9)}$$

as indicated in Example 2.3.

2. Let

$$f(t) = \begin{cases} \sin(3t) & \text{for } -\pi \leq t \leq \pi \\ 0 & \text{otherwise.} \end{cases}$$

Compute $\widehat{f}(\lambda)$ (i.e., provide details of Example 2.4).

3. (*Note*: A computer algebra system would help for this problem). Let

$$f(t) = \begin{cases} t^2 + 4t + 4 & \text{for } -2 \leq t \leq -1 \\ 2 - t^2 & \text{for } -1 \leq t \leq 1 \\ t^2 - 4t + 4 & \text{for } 1 \leq t \leq 2 \\ 0 & \text{otherwise.} \end{cases}$$

Show that f and f' are continuous everywhere (pay particular attention to what happens at $t = -2$, -1, 1, and 2). Draw a careful graph of f. Compute its Fourier transform and show that

$$\widehat{f}(\lambda) = \left(\frac{8\sin(\lambda)(1 - \cos(\lambda))}{\sqrt{2\pi}\lambda^3}\right).$$

Plot the Fourier transform over the interval $-10 \leq \lambda \leq 10$. Note that the Fourier transform decays like $\frac{1}{\lambda^3}$ as $\lambda \mapsto \infty$. Also note that in Exercise 1, the given function is discontinuous and $\widehat{f}(\lambda)$ decays like $\frac{1}{\lambda}$ as $\lambda \mapsto \infty$; in Exercise 2, f is continuous but not differentiable and $\widehat{f}(\lambda)$ decays like $1/\lambda^2$. Do you see a pattern?

4. Suppose $f \in L^2(-\infty, \infty)$ is a real-valued, even function; show that $\widehat{f}$ is real valued. Suppose $f \in L^2(-\infty, \infty)$ is a real-valued, odd function; show that $\widehat{f}(\lambda)$ is purely imaginary (its real part is zero) for each λ.

5. Let

$$\phi(x) = \begin{cases} 1 & \text{if } 0 \le x < 1 \\ 0 & \text{otherwise.} \end{cases}$$

Show that

$$(\phi * \phi)(x) = \begin{cases} 1 - |x - 1| & \text{if } 0 \le x < 2 \\ 0 & \text{otherwise.} \end{cases}$$

6. Let $f_s(x) = \sqrt{s}e^{-sx^2}$. Show that

$$\hat{f}_s(\lambda) = \frac{1}{\sqrt{2}}e^{\frac{-\lambda^2}{4s}}.$$

Hint: After writing out the definition of the Fourier transform, complete the square in the exponent; then perform a change of variables in the resulting integral (a complex translation) and then use the fact that $\int_{-\infty}^{\infty} e^{-x^2}\, dx = \sqrt{\pi}$.

7. Establish the parts of Theorem 2.6 that deal with the inverse Fourier transform. Establish the relationship between the Fourier transform and the Laplace transform given in the last part of this theorem.

8. Suppose L is a time-invariant filter with response function h (see Theorem 2.19). Show that if L is causal, then $h(t) = 0$ for $t < 0$. *Hint:* Prove by contradiction—assume $h(t_0) \ne 0$ for some $t_0 < 0$; show that by choosing an appropriate f that is nonzero only on $[0, \delta]$ for some very small $\delta > 0$, $L(f)(t) \ne 0$ for some $t < 0$ even though $f(t) = 0$ for all $t < 0$ (thus contradicting causality).

9. Using integration by parts, show that (2.27) follows from (2.26).

10. Let

$$h(t) = \begin{cases} Ae^{-\alpha t} & \text{if } t \ge 0 \\ 0 & \text{if } t < 0 \end{cases}$$

where A and α are parameters as in Example 2.21. Show that

$$\hat{h}(\lambda) = \frac{1}{\sqrt{2\pi}}(\mathcal{L}h)(i\lambda) = \frac{A}{\sqrt{2\pi}(\alpha + i\lambda)}.$$

11. Consider the signal

$$f(t) = e^{-t}\left(\sin 5t + \sin 3t + \sin t + \sin 40t\right) \quad 0 \le t \le \pi.$$

Filter this signal with the Butterworth filter [i.e., compute $(f * h)(t)$ for $0 \le t \le \pi$]. Try various values of $A = a$ (starting with $A = a = 10$). Compare the filtered signal with the original signal.

12. Consider the filter given by $f \mapsto f * h$, where

$$h(s) := \begin{cases} 1/d & 0 \le t \le d \\ 0 & \text{otherwise.} \end{cases}$$

Compute and graph $\widehat{h}(\lambda)$ for $0 \le \lambda \le 20$ for various values of d. Suppose you wish to use this filter to remove signal noise with frequency larger than 30 and retain frequencies in the range of 0 to 5 (cycles per 2π-interval). What values of d would you use? Filter the signal

$$f(t) = e^{-t} \left(\sin 5t + \sin 3t + \sin t + \sin 40t \right) \quad 0 \le t \le \pi$$

with your chosen values of d and compare the graph of the filtered signal with the graph of f.

13. The sampling theorem (Theorem 2.23) states that if the Fourier transform of a signal f is band limited to the interval $[-\Omega, \Omega]$, then the signal can be recovered by sampling the signal at a grid of time nodes separated by a time interval of π/Ω. Now suppose that the Fourier transform of a signal, f, vanishes outside the interval $[\omega_1 \le \lambda \le \omega_2]$; develop a formula analogous to (2.21) where the time interval of separation is $2\pi/(\omega_1 + \omega_2)$. *Hint*: Show that the Fourier transform of the signal

$$g(t) = e^{-it\left(\frac{\omega_1 + \omega_2}{2}\right)} f(t)$$

is band limited to the inteval $[-\Omega, \Omega]$ with $\Omega = (\omega_2 - \omega_1)/2$ and then apply Theorem 2.23 to the signal g.

14. (*Oversampling*) This exercise develops a version of the sampling theorem which converges faster than the version given in Theorem 2.23. Complete the following outline.

   (a) Suppose f is a band-limited signal with $\widehat{f}(\lambda) = 0$ for $|\lambda| \ge \Omega$. Fix a number $a > 1$. Repeat the proof of Theorem 2.23 to show that

$$\widehat{f}(\lambda) = \sum_{n=-\infty}^{\infty} c_{-n} e^{-in\pi\lambda/a\Omega} \quad \text{with } c_{-n} = \frac{\pi}{\sqrt{2\pi a\Omega}} f\left(\frac{n\pi}{a\Omega}\right).$$

   (b) Let $\widehat{g_a}(\lambda)$ be the function whose graph is given by Figure 21. Show that

$$g_a(t) = \frac{\sqrt{2}(\cos(\Omega t) - \cos(a\Omega t))}{\sqrt{\pi}(a-1)\Omega t^2}.$$

   (c) Since $\widehat{f}(\lambda) = 0$ for $|\lambda| \ge \Omega$, $\widehat{f}(\lambda) = \widehat{f}(\lambda)\widehat{g_a}(\lambda)$. Use Theorem 2.1, Theorem 2.6, and the expressions for $\widehat{f}$ and g_a in parts (a) and (b) to show that

$$f(t) = \sum_{n=-\infty}^{\infty} \frac{\pi}{\sqrt{2\pi} a\Omega} f\left(\frac{n\pi}{a\Omega}\right) g_a\left(t - \frac{n\pi}{a\Omega}\right). \tag{2.28}$$

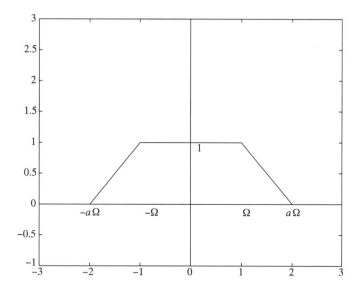

Figure 21 Graph of $\widehat{g}_a$

Since $g_a(t)$ has a factor of t^2 in the denominator, this expression for $f(t)$ converges faster than the expression for f given in Theorem 2.23 (the nth term behaves like $1/n^2$ instead of $1/n$). The disadvantage of (2.28) is that the function is sampled on a grid of $n\pi/(a\Omega)$, which is a more frequent rate of sampling than the grid $n\pi/\Omega$ (since $a > 1$) used in Theorem 2.23. Thus there is a trade-off between the sample rate and the rate of convergence.

Chapter 3

Discrete Fourier Analysis

The Fourier transform and Fourier series techniques are useful for analyzing continuous signals such as the graph in Figure 1. However, for many applications, the signal is a discrete data set (see Figure 2), such as the signal coming from a compact disc player. A discrete version of the Fourier transform is needed to analyze discrete signals.

3.1   The Discrete Fourier Transform

To motivate the idea behind the discrete Fourier transform, we numerically approximate the coefficients of a Fourier series for a continuous function $f(t)$. The trapezoidal rule for approximating the integral $(2\pi)^{-1} \int_0^{2\pi} F(t)\, dt$ with step size $h = 2\pi/n$ is

$$(2\pi)^{-1} \int_0^{2\pi} F(t)\, dt \approx \frac{1}{2\pi} \frac{2\pi}{n} \left[\frac{Y_0}{2} + Y_1 + \cdots + Y_{n-1} + \frac{Y_n}{2} \right]$$

where $Y_j := F(hj) = F(2\pi j/n)$, $j = 0 \ldots n$. If $F(t)$ is 2π-periodic, then $Y_0 = Y_n$ and the preceding formula becomes

$$(2\pi)^{-1} \int_0^{2\pi} F(t)dt \approx \frac{1}{n} \sum_{j=0}^{n-1} Y_j.$$

Applying this formula to the computation of the kth complex Fourier coefficient gives

$$\alpha_k = \frac{1}{2\pi} \int_0^{2\pi} f(t)\exp(-ikt)\, dt \approx \frac{1}{n} \sum_{j=0}^{n-1} f(2\pi j/n)\exp(-2\pi ijk/n).$$

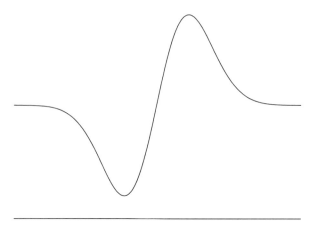

Figure 1 Continuous signal

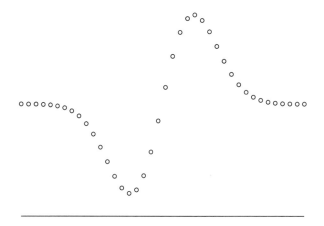

Figure 2 Discrete signal

Therefore,

$$\alpha_k \approx \frac{1}{n} \sum_{j=0}^{n-1} y_j \overline{w}^{jk}, \tag{3.1}$$

where

$$y_j = f(2\pi j/n) \quad \text{and} \quad w = \exp(2\pi i/n).$$

The sum on the right side of equation (3.1) involves y_j, which is the value of f at $t = 2\pi j/n$. The values of f at the other t-values are not needed. This sum will be used in the next section as the definition of the Fourier transform of a discrete signal, whose values may only be known at a discrete set of time nodes $t = 2\pi j/n$ for $j = 0, \ldots, n-1$.

The right side of equation (3.1) is unchanged if k is replaced by $k + n$ since $\overline{w}^n = e^{-2\pi i} = 1$. Thus, the right side of (3.1) does not approximate α_k for $k \geq n$, because the α_k are not n-periodic. In fact, this expression only approximates α_k for k that are relatively small compared with n because the trapezoidal rule algorithm only provides accurate numerical approximations if the step size, $h = 2\pi/n$, is small relative to the frequency k.

3.1.1 Definition of Discrete Fourier Transform

Let $\mathcal{S}_n$ be the set of n-periodic sequences of complex numbers. Each element, $y = \{y_j\}_{j=-\infty}^{\infty}$ in $\mathcal{S}_n$, can be thought of as a periodic discrete signal where y_j is the value of the signal at a time node $t = t_j$. The sequence y_j is n-periodic if $y_{k+n} = y_k$ for any integer k. The set $\mathcal{S}_n$ forms a complex vector space under the operations of entry-by-entry addition and entry-by-entry multiplication by a scalar. If $x = \{x_j\}_{j=-\infty}^{\infty} \in \mathcal{S}_n$ and $y = \{y_j\}_{j=-\infty}^{\infty} \in \mathcal{S}_n$, then the jth component of $\{x + y\}$ is $x_j + y_j$ and the jth component of $c\{x\}$ is cx_j. Here, n should be thought of as the number of time nodes corresponding to the discrete signal of interest.

DEFINITION 3.1   *Let $y = \{y_j\}_{j=-\infty}^{\infty} \in \mathcal{S}_n$. The discrete Fourier transform of y is the sequence $(\mathcal{F}_n\{y\})_k = \widehat{y}_k$, where*

$$\widehat{y}_k = \sum_{j=0}^{n-1} y_j \overline{w}^{jk} \quad with \quad w = \exp(2\pi i/n).$$

In detail, the discrete Fourier transform is the sequence

$$\widehat{y}_k = \sum_{j=0}^{n-1} y_j \exp(-2\pi i k j/n).$$

The formula for the discrete Fourier transform is analogous to the formula for the kth Fourier coefficient with the sum over j taking the place of the integral over t (see Theorem 1.18). As stated in the previous section, if the y_j arise as values from a continuous signal, f, defined on the interval $[0, 2\pi]$, then the kth Fourier coefficient of $f(\alpha_k)$ is approximated by

$$\alpha_k \approx \frac{1}{n}\widehat{y}_k$$

for k small relative to n.

The computation of the discrete Fourier transform is equivalent to the following matrix computation:

$$\mathcal{F}_n\{y\} = \widehat{y} = (\overline{F}_n) \cdot (y)$$

where $y = (y_0, \ldots, y_{n-1})^T$ and $\widehat{y} = (\widehat{y}_0, \ldots, \widehat{y}_{n-1})^T$ and where

$$
F_n = \begin{pmatrix}
1 & 1 & 1 & \cdots & 1 \\
1 & w & w^2 & \cdots & w^{n-1} \\
1 & w^2 & w^4 & \cdots & w^{2(n-1)} \\
\vdots & & \ddots & & \vdots \\
1 & w^{n-1} & w^{2(n-1)} & \cdots & w^{(n-1)^2}
\end{pmatrix}. \tag{3.2}
$$

3.1.2   Properties of the Discrete Fourier Transform

The discrete Fourier transform of a signal in $\mathcal{S}_n$ produces a discrete signal that also belongs to $\mathcal{S}_n$, as the following lemma shows.

LEMMA 3.2 $\mathcal{F}_n$ *is a linear operator from* $\mathcal{S}_n$ *to* $\mathcal{S}_n$.

Proof   To show that $\mathcal{F}_n\{y\}$ belongs to $\mathcal{S}_n$ we must show that $\mathcal{F}_n\{y\}_k$ is n-periodic. We have

$$
\widehat{y}_{k+n} = \sum_{j=0}^{n-1} y_j \overline{w}^{j(k+n)}
$$

$$
= \sum_{j=0}^{n-1} y_j \overline{w}^{jk} \overline{w}^{nj}
$$

$$
= \sum_{j=0}^{n-1} y_j \overline{w}^{jk} \quad [\text{because } \overline{w}^n = e^{-(2\pi i/n)n} = 1]
$$

$$
= \widehat{y}_k.
$$

Therefore, $\widehat{y}_{k+n} = \widehat{y}_k$ and so this sequence is n-periodic. Finally, since matrix multiplication is a linear process, $\mathcal{F}_n$ is a linear operator.   ♦

Our next task is to compute the inverse of the discrete Fourier transform. We have already computed the inverse Fourier transform since Theorem 2.1 tells us how to recover the function f from its Fourier transform. The inverse discrete Fourier transform is analogous; it allows us to recover the original discrete signal, y, from its discrete Fourier transform, $\widehat{y} = \overline{F}_n(y)$. Computing the inverse discrete Fourier transform is equivalent to computing the inverse of the matrix $\overline{F}_n$. The following theorem gives an easy formula for the inverse of this matrix.

THEOREM 3.3   *Suppose* $y = \{y_k\}$ *is an element of* $\mathcal{S}_n$. *Let* $\mathcal{F}_n(y) = \widehat{y}$; *that is,*

$$
\widehat{y}_k = \sum_{j=0}^{n-1} y_j \overline{w}^{jk}
$$

where $w = \exp(2\pi i/n)$. Then $y = \mathcal{F}_n^{-1}(\widehat{y})$ is given by

$$y_j = \frac{1}{n} \sum_{k=0}^{n-1} \widehat{y}_k w^{jk}.$$

Remark. Using matrix terminology, Theorem 3.3 asserts that the inverse discrete Fourier transform is given by matrix multiplication with the matrix $\frac{1}{n} F_n$ [F_n is defined in (3.2)]; that is,

$$y = \mathcal{F}_n^{-1} \widehat{y} = \frac{1}{n}(F_n) \cdot (\widehat{y}).$$

Since $\widehat{y} = (\overline{F}_n)(y)$, the equation $y = \frac{1}{n} F_n(\widehat{y})$ is equivalent to

$$I_n = \left(\frac{F_n}{\sqrt{n}}\right)\left(\frac{\overline{F}_n}{\sqrt{n}}\right) \qquad \text{where } I_n = n \times n \text{ identity matrix.} \tag{3.3}$$

Since F_n is symmetric, this equation means that the matrix $F_n/\sqrt{n}$ is unitary (its inverse equals the conjugate of its transpose).

Proof To establish equation (3.3), we must show that

$$\frac{1}{n} \sum_{k=0}^{n-1} w^{\ell k} \, \overline{w}^{kj} = \begin{cases} 1 & \text{if } j = \ell \\ 0 & \text{otherwise.} \end{cases}$$

Since $w = e^{2\pi i/n}$, clearly $\overline{w} = w^{-1}$ and so

$$w^{\ell k} \overline{w}^{kj} = w^{k(\ell - j)}.$$

To sum this expression over $k = 0, \ldots, n-1$, we use the following identity [see equation (1.23) in the proof of Lemma 1.23]

$$\sum_{k=0}^{n-1} z^k = \begin{cases} n & \text{if } z = 1 \\ \dfrac{1 - z^n}{1 - z} & \text{if } z \neq 1. \end{cases}$$

Set $z = w^{\ell - j}$; note that $z^n = 1$ because $w^n = 1$. Also note that $w^{\ell - j} \neq 1$ unless $j = \ell$ for $0 \leq j, \ell \leq n-1$. The previous equation becomes

$$\frac{1}{n} \sum_{k=0}^{n-1} w^{(\ell - j)k} = \begin{cases} 1 & \text{if } j = \ell \\ 0 & \text{if } j \neq \ell. \end{cases}$$

Thus

$$\frac{1}{n} \sum_{k=0}^{n-1} w^{\ell k} \overline{w}^{kj} = \frac{1}{n} \sum_{k=0}^{n-1} w^{(\ell - j)k} = \begin{cases} 1 & \text{if } j = \ell \\ 0 & \text{otherwise} \end{cases}$$

as desired. ◆

Both $\mathcal{F}_n$ and $\mathcal{F}_n^{-1}$ are linear transformations from $\mathcal{S}_n$ to itself. Here are some additional properties that are analogous to the corresponding properties for the Fourier transform given in Theorems 2.6 and 2.10. The derivations of these properties will be left as exercises.

THEOREM 3.4 *The following properties hold for the discrete Fourier transform.*

- *Shifts or translations. If $y \in \mathcal{S}_n$ and $z_k = y_{k+1}$, then $\mathcal{F}[z]_j = w^j \mathcal{F}[y]_j$, where $w = e^{2\pi i/n}$.*

- *Convolutions. If $y \in \mathcal{S}_n$ and $z \in \mathcal{S}_n$, then the sequence defined by*

$$[y * z]_k := \sum_{j=0}^{n-1} y_j z_{k-j}$$

  is also in $\mathcal{S}_n$. The sequence $y * z$ is called the convolution of the sequences y and z.

- *The Convolution theorem: $\mathcal{F}[y * z]_k = \mathcal{F}[y]_k \mathcal{F}[z]_k$.*

- *If $y \in \mathcal{S}_n$ is a sequence of real numbers, then*

$$\mathcal{F}[y]_{n-k} = \overline{\mathcal{F}[y]}_k \quad \text{for } 0 \leq k \leq n$$
$$\text{or} \quad \widehat{y}_{n-k} = \overline{\widehat{y}}_k.$$

Since $\widehat{y}_{n-k} = \overline{\widehat{y}}_k$, only the $\widehat{y}_0, \ldots, \widehat{y}_{n/2-1}$ need to be calculated (assuming n is even). Further savings in calculations are possible by using the fast Fourier transform algorithm, which we discuss in the next section.

At the end of Section 3.1, we indicated that if $y_k = f(k/2\pi n)$ for a continuous function, f, then $\widehat{y}_k$ is an accurate approximation of the kth Fourier coefficient of f, α_k, when k is small relative to n. The equation $\widehat{y}_{n-k} = \overline{\widehat{y}}_k$ further emphasizes this point. For example, if n is large and $k = 1$, then $\overline{\widehat{y}}_{n-1} = \widehat{y}_1$, which approximates α_1—a low-frequency coefficient. Therefore, $\widehat{y}_{n-1}$ has no resemblance to α_{n-1}—a high-frequency coefficient.

3.1.3   The Fast Fourier Transform

Since the discrete Fourier transform is such a valuable tool in signal analysis, an efficient algorithm for computing the discrete Fourier and inverse Fourier transforms is very desirable. The computation of the discrete Fourier transform is equivalent to multiplying the sequence y (written as an $n \times 1$ column vector) by the $n \times n$ matrix $\overline{F}_n$—an operation that requires n^2 multiplications. However, the fast Fourier transform (FFT) is an algorithm that takes advantage of the special form of the matrix F_n to reduce the number of multiplications to roughly $5n \log_2 n$. If $n = 1000$, then this simplification reduces the number of multiplications from a million to about 50,000. The relative savings increase as n gets larger.

The fast Fourier transform algorithm will require an even number of nodes $n = 2N$. We consider a sequence $y = (y_0, \ldots, y_{2N-1})$ (extended periodically, with period $n = 2N$). The $\hat{y}_k$ are calculated via

$$\hat{y}_k = \sum_{j=0}^{2N-1} y_j \overline{w}^{jk}.$$

Splitting the preceding sum into even and odd indices yields

$$\hat{y}_k = \sum_{j=0}^{N-1} y_{2j} \overline{w}^{2jk} + \sum_{j=0}^{N-1} y_{2j+1} \overline{w}^{(2j+1)k}.$$

Recall that $w = e^{2\pi i/n}$ and $n = 2N$. Let $W = \exp(2\pi i/N) = w^2$. We obtain

$$\hat{y}_k = \sum_{j=0}^{N-1} y_{2j} \bar{W}^{jk} + \overline{w}^k \left(\sum_{j=0}^{N-1} y_{2j+1} \bar{W}^{jk} \right).$$

The preceding equation expresses $\hat{y}_k$ in terms of two discrete Fourier transforms with n replaced by N:

$$\hat{y}_k = \mathcal{F}_N[\{y_0, y_2, \cdots, y_{2N-2}\}]_k + \overline{w}^k \mathcal{F}_N[\{y_1, y_3, \cdots, y_{2N-1}\}]_k$$

for $0 \le k \le 2N - 1$. Further savings are possible. In the last equation, replace k by $k + N$ and use the following facts:

- $\mathcal{F}_N[y_{\text{even}}]$ and $\mathcal{F}_N[y_{\text{odd}}]$ both have period N.
- $\overline{w}^{k+N} = \overline{w}^k \exp(-\pi i) = -\overline{w}^k$.

The result is that for $0 \le k \le N - 1$ we have

$$\hat{y}_k = \mathcal{F}_N[\{y_0, y_2, \ldots, y_{2N-2}\}]_k + \overline{w}^k \mathcal{F}_N[\{y_1, y_3, \ldots, y_{2N-1}\}]_k \qquad (3.4)$$

$$\hat{y}_{k+N} = \mathcal{F}_N[\{y_0, y_2, \ldots, y_{2N-2}\}]_k - \overline{w}^k \mathcal{F}_N[\{y_1, y_3, \ldots, y_{2N-1}\}]_k. \qquad (3.5)$$

Thus we have described $(\mathcal{F}_{2N} y)_k = \hat{y}_k$ for $0 \le k \le 2N - 1$ in terms of $\mathcal{F}_N[\{y_0, y_2, \ldots, y_{2N-2}\}]_k$ and $\mathcal{F}_N[\{y_1, y_3, \ldots, y_{2N-1}\}]_k$ for $0 \le k \le N - 1$. Similar formulas can be derived for the inverse discrete Fourier transform; they are

$$y_k = \frac{1}{2} \left\{ \mathcal{F}_N^{-1}[\{\hat{y}_0, \hat{y}_2, \ldots, \hat{y}_{2N-2}\}]_k + w^k \mathcal{F}_N^{-1}[\{\hat{y}_1, \hat{y}_3, \ldots, \hat{y}_{2N-1}\}]_k \right\}$$

$$y_{k+N} = \frac{1}{2} \left\{ \mathcal{F}_N^{-1}[\{\hat{y}_0, \hat{y}_2, \ldots, \hat{y}_{2N-2}\}]_k - w^k \mathcal{F}_N^{-1}[\{\hat{y}_1, \hat{y}_3, \ldots, \hat{y}_{2N-1}\}]_k \right\}$$

for $0 \le k \le N-1$ (the factor of $\frac{1}{2}$ appears because the inversion formula contains a factor of $1/n$).

In matrix terms the computation of the discrete Fourier transform is given by

$$\mathcal{F}_{2N}y = (\overline{F}_{2N})(y) = \begin{pmatrix} I_N & \overline{D}_N \\ I_N & -\overline{D}_N \end{pmatrix} \begin{pmatrix} \overline{F}_N & 0 \\ 0 & \overline{F}_N \end{pmatrix} \begin{pmatrix} y_{\text{even}} \\ y_{\text{odd}} \end{pmatrix}$$

where F_N is the matrix defined in (3.2), I_N is the $N \times N$ identity matrix, and D_N is the diagonal matrix with the entries $1, w, w^2, \ldots, w^{N-1}$ on the diagonal.

The number of multiplications required to compute the discrete Fourier transform from its definition is about $n^2 = 4N^2$ [since the discrete Fourier transform requires multiplication by the $n \times n$ matrix F_n given in (3.2)]. The number of operations required to compute $\hat{y}_k$ and $\hat{y}_{k+N}$ in equations (3.4) and (3.5) is $2N^2 + N + 1$ (since $\mathcal{F}_N[\{y_0, y_2, \ldots, y_{2N-2}\}]_k$ and $\mathcal{F}_N[\{y_1, y_3, \ldots, y_{2N-1}\}]_k$ both require N^2 multiplications). Therefore, the number of multiplications has almost been cut in half (provided N is large). If N is a multiple of 2, then the process can be repeated by writing $\mathcal{F}_N$ as a product involving $\mathcal{F}_{N/2}$. This would cut the number of multiplications by almost another factor of $\frac{1}{2}$ or almost one-quarter of the original number.

To keep going, we assume the original value of n is a power of 2, say $n = 2^L$. Then we can iterate the preceding procedure L times. The Lth iteration involves multiplication by the 1×1 matrix $\overline{F}_1$, which is just the scalar 1. How many multiplications are involved with the full implementation of this algorithm? The answer is found as follows. Let K_L be the number of real multiplications required to compute $\mathcal{F}[y]$ by the preceding method for $n = 2^L$ (and so $N = 2^{L-1}$). From the formulas derived previously, we see that to compute $\mathcal{F}[y]$, we need to compute $\mathcal{F}[y_{\text{even}}]$ and $\mathcal{F}[y_{\text{odd}}]$. The number of multiplications required is proportional to $2K_{L-1} + N = 2K_{L-1} + 2^{L-1}$. Thus, K_L is related to K_{L-1} via

$$K_L \approx 2K_{L-1} + 2^{L-1}.$$

When $L = 0$, $n = 2^0 = 1$ and no multiplications are required; thus, $K_0 = 0$. Inserting $L = 1$ in the last equation, we find that $K_1 \approx 1$. Similarly, setting $L = 2$ then yields $K_2 \approx 2 \times 2^1$. Similarly, $K_3 \approx 3 \times 2^2$, $K_4 \approx 4 \times 2^3$, and so on. The general formula is $K_L \approx L \times 2^{L-1}$. Setting $n = 2^L$ and noting that $L = \log_2 n$, we see that the number of multiplications required is proportional to $n \log_2 n$. The actual proportionality constant is about 5.

Similar algorithms can be obtained when $n = N_1 N_2 \cdots N_m$, although the fastest one is obtained in the case discussed previously. For a discussion of this and related topics, consult the references.

EXAMPLE 3.5

A plot of the absolute value of the discrete Fourier coefficients for the function $f(t) = t + t^2$ is given in Figure 3 (this graph was produced by MATLAB's fft routine). The horizontal axis is $k = 0, 1, 2, \ldots, n$, with $n = 64$, and the dots represent $(k, |\hat{y}_k|)$. Since the Fourier coefficients decay with frequency (by the Riemann-Lebesgue lemma), the $|\hat{y}_k|$ decrease with k until $k = n/2$. The graph of the $|\hat{y}_k|$ is symmetric about the line $k = n/2 = 32$ since $|\hat{y}_{n-k}| = |\hat{y}_k|$ (so the values of $|\hat{y}_{n/2+1}|, \ldots |\hat{y}_{n-1}|$ are the same as $|\hat{y}_{n/2-1}|, \ldots, |\hat{y}_1|$). ∎

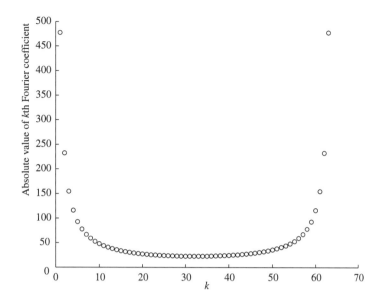

Figure 3 Plot of fast Fourier transform coefficients for $f = t + t^2$

EXAMPLE 3.6

We consider the signal in Figure 4, which is generated by the function

$$y(t) = e^{-(\cos t)^2}(\sin 2t + 2\cos 4t + 0.4\sin t \sin 50t)$$

over the interval $0 \le t \le 2\pi$. We wish to filter the high-frequency noise from this signal. The signal is first discretized by evaluating it at $2^8 = 256$ equally spaced points on the interval $0 \le t \le 2\pi$. Then the fast Fourier transform is used to generate the discrete Fourier coefficients $\widehat{y}_k$, $k = 0, \ldots, 255$. Judging from Figure 4, the noise has a frequency that is larger than 5 (cycles per 2π-interval). Thus we keep only $\widehat{y}_k$ for $0 \le k \le 5$ and set $\widehat{y}_k = 0$ for $6 \le k \le 128$. By Theorem 3.4, $\widehat{y}_k = 0$ for $128 \le k \le 250$. Applying the inverse fast Fourier transform to the filtered $\widehat{y}_k$, we assemble the filtered signal whose graph is given in Figure 5.   ∎

EXAMPLE 3.7

We consider the signal given in Figure 6, which is generated by the function

$$y(t) = e^{-t^2/10}(\sin 2t + 2\cos 4t + 0.4\sin t \sin 10t)$$

over the interval $0 \le t \le 2\pi$. We wish to compress this signal by taking its fast Fourier transform and ignoring the small Fourier coefficients. As in the previous example, we sample the signal at $2^8 = 256$ equally spaced nodes. We apply the fast Fourier transform to generate $\widehat{y}_k$ and set 80% of the $\widehat{y}_k$ equal to

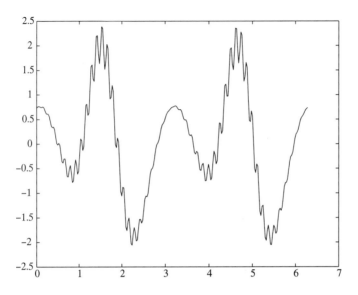

Figure 4 Unfiltered signal

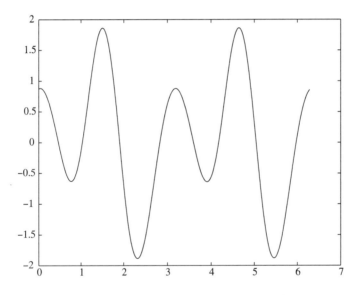

Figure 5 Filtered signal with FFT, keeping only $\widehat{f}_k$, $k = 0, \ldots, 5$

zero (the smallest 80%). Taking the inverse fast Fourier transform of the new $\widehat{y}_k$, we assemble the compressed signal that is graphed in Figure 7. Notice the Gibbs effect since the partial Fourier series is periodic, and so its values at 0 and 2π are the same.

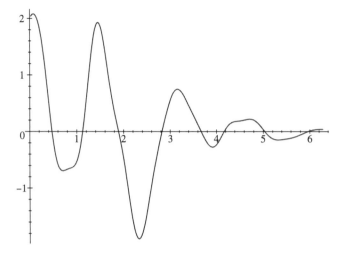

Figure 6 Uncompressed signal

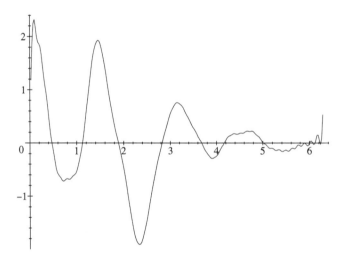

Figure 7 Eighty percent compressed signal with FFT

The relative error between the original signal y and the compressed signal yc can be computed to be

$$\text{Relative error} = \frac{\|y - yc\|_{l^2}}{\|y\|_{l_2}} = 0.1018$$

(see Appendix B on MATLAB code to learn how to compute the relative error using MATLAB).   ∎

3.1.4   The FFT Approximation to the Fourier Transform

A very important application of the FFT is the computation of a numerical approximation to the Fourier transform of a function f defined on an interval $a \le t \le b$ from samples taken at n nodes, spaced a constant interval T units apart [i.e., at $a, a+T, a+2T, \ldots, a+(n-1)T = b-T$, where $T = (b-a)/n$]. We assume that the function f is zero outside of the interval $[a, b]$, continuous on $a \le t < b$, and that it satisfies $f(b) = f(a)$. Since most signals are not periodic, this last assumption will usually result in a discontinuity in the signal at $t = b$, thus causing a Gibbs phenomenon when the function is reconstructed from the Fourier transform of the data.

The Fourier transform of f is given by

$$\hat{f}(\omega) = \frac{1}{\sqrt{2\pi}} \int_a^b f(t) e^{-i\omega t}\, dt.$$

We want to change variables so that the time interval runs from 0 to 2π rather than a to b. Letting $\theta = 2\pi \frac{t-a}{b-a}$ (or $t = a + (b-a)\theta/2\pi$), $\hat{f}$ is then given by

$$\hat{f}(\omega) = \frac{b-a}{(2\pi)^{3/2}} \int_0^{2\pi} f(a + (b-a)\theta/(2\pi)) e^{-i\omega(a+(b-a)\theta/(2\pi))}\, d\theta$$

$$= \frac{b-a}{(2\pi)^{3/2}} e^{-i\omega a} \int_0^{2\pi} f(a + (b-a)\theta/(2\pi)) e^{-i\frac{(b-a)\omega}{2\pi}\theta}\, d\theta.$$

Note that if we define the function g and the frequency ω_k via

$$g(\theta) := f(a + (b-a)\theta/(2\pi)) \quad \text{and} \quad \omega_k = \frac{2\pi}{b-a}k,$$

then $\hat{f}(\omega_k)$ has the form

$$\hat{f}(\omega_k) = \frac{b-a}{\sqrt{2\pi}} e^{-i\omega_k a} \left\{ \frac{1}{2\pi} \int_0^{2\pi} g(\theta) e^{-ik\theta}\, d\theta \right\} = \frac{b-a}{\sqrt{2\pi}} e^{-i\omega_k a} \hat{g}_k$$

where $\hat{g}_k$ is the kth Fourier series coefficient for g.

The relationship between the discrete Fourier transform (DFT) and Fourier series coefficients is one that we have dealt with before. First, let

$$y_j := g\left(j \underbrace{\frac{2\pi}{n}}_{h} \right) = f\left(a + j \underbrace{\frac{b-a}{n}}_{T} \right), \quad j = 0, \ldots, n-1.$$

Thus the y_j's are the samples of f taken at times $a + jT$. The DFT approximation to $\hat{g}_k$ is $\hat{y}_k/n$. This implies that the DFT for the y_j's is related to $\hat{f}$ via

$$\hat{f}(\omega_k) \approx \frac{b-a}{n\sqrt{2\pi}} e^{-i\omega_k a} \hat{y}_k = \frac{T}{\sqrt{2\pi}} e^{-i\omega_k a} \hat{y}_k. \tag{3.6}$$

Formula (3.6) accomplishes our goal of approximating the Fourier transform of f at the frequencies ω_k by the values of f at n equally spaced nodes on $a \le t \le b$.

3.1.5   Application—Parameter Identification

As an application of the discrete Fourier transform, we consider the problem of predicting the sway of a building resulting from wind or other external forces. Buildings, especially very tall ones, sway and vibrate in the wind. This is common and hard to avoid. Too much sway and vibration can cause structural damage. How do we find out how a structure behaves in a strong wind? A rough model for the horizontal displacement u due to a wind exerting a force f is

$$au'' + bu' + cu = f(t) \tag{3.7}$$

where a, b, and c are positive constants, with $b^2 - 4ac < 0$. This is a standard problem in ordinary differential equations. The twist here is that we don't know what a, b, and c are. Instead, we have to *identify* them experimentally.

One way to do this is to hit a wall with a jackhammer and record the response. The impulse force exerted by the jackhammer is modeled by $f(t) = f_0\delta(t)$, where $\delta(t)$ is Dirac's δ-function, and f_0 is the strength of the impulse. Think of the δ-function as a signal, with total strength equal to one and that lasts for a very short time. Technically speaking, the δ-function is the limit as $h \mapsto 0$ of f_h whose graph is given in Figure 8. The corresponding impulse response u has the form (see [3, §6.4])

$$u(t) = \begin{cases} 0 & t < 0 \\ \dfrac{f_0}{wa}\sin(\omega t)e^{-\mu t} & t \geq 0 \end{cases} \tag{3.8}$$

where $\omega = \dfrac{\sqrt{4ac - b^2}}{2a}$ and $\mu = \dfrac{b}{2a}$.

The idea now is to take measurements of the the displacements (or sway) at various intervals after the building is hit by the jackhammer. This will give

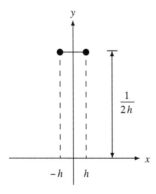

Figure 8 Graph of f_h

a (discretized) graph of the function u. In theory, the frequency ω can be determined from this graph along with decay constant μ and the amplitude $f_0/(\omega a)$. Since f_0 is known (the strength of the jackhammer), the parameter a can be determined from the amplitude $f_0/(\omega a)$ and ω. The parameters b and c can then be determined by the equations $\omega = \dfrac{\sqrt{4ac - b^2}}{2a}$ and $\mu = \dfrac{b}{2a}$ (two equations and two unknowns).

The trouble with this approach is that the actual data (the measurements of the displacements) will look somewhat different from the theoretical response u in (3.8). The measured data will contain noise that adds high frequencies to the response. The linear differential equation and the δ-function used in the preceding model are only rough approximations to the corresponding physical system and actual applied force of the jackhammer. Small nonlinearities in the system and an impulse lasting a finite amount of time will add oscillations to the response. Thus, the frequency ω may be difficult to infer directly from the graph of the data.

In spite of all of the additional factors mentioned previously, an approximate value of ω can still be found. This is where the discrete Fourier transform plays a key role. The frequency ω will be the largest-frequency component present in the data, and this largest frequency can be found easily by computing the discrete Fourier transform of the data. An approximate value for the decay constant μ can also be determined by analyzing the amplitude of the data. Then approximate values for the parameters a, b, and c can be determined as indicated previously. These ideas are further investigated in Exercises 3–6.

3.1.6 Application—Discretizations of Ordinary Differential Equations

As another application of the DFT, we consider the differential equation in equation (3.7) and take f to be a continuous, 2π-periodic function of t. There is a well-known analytical method for finding the unique periodic solution to this equation (cf. [3, §3.7.2]), provided that f is known for all t. On the other hand, if we only know f at the points $t_j = 2\pi j/n$, for some integer $n \geq 1$, this method is no longer applicable.

Instead of trying to work directly with the differential equation itself, we work with a discretized version of it. There a many ways of discretizing. With $h = 2\pi/n$, the one that we use here involves the following approximations:

$$u'(t) \approx \frac{u(t) - u(t - h)}{h},$$

$$u''(t) \approx \frac{u(t + h) + u(t - h) - 2u(t)}{h^2}.$$

We let $t_k = 2\pi k/n$ and $u_k = u(t_k)$ for $0 \leq k \leq n$. Since u is periodic, $u_n = u_0$.

At $t = t_k$, the preceding difference formulas become

$$u'(t_k) \approx \frac{u_k - u_{k-1}}{h}, \tag{3.9}$$

$$u''(t_k) \approx \frac{u_{k+1} + u_{k-1} - 2u_k}{h^2}. \tag{3.10}$$

for $1 \le k \le n - 1$. Substituting these values into the differential equation (3.7) and collecting terms yields the following difference equation (see Exercise 10):

$$au_{k+1} + \beta u_k + \gamma\, u_{k-1} = h^2 f_k, \quad 1 \le k \le n - 1 \tag{3.11}$$

where $f_k = f(2\pi k/n)$, $\beta = ch^2 + bh - 2a$, and $\gamma = a - bh$.

Suppose that $u \in \mathcal{S}_n$ is a solution to the difference equation (3.11). Let $\hat{u} = \mathcal{F}[u]$ and $\hat{f} = \mathcal{F}[f]$ and $w = \exp(2\pi i/n)$. By taking the DFT of both sides and using the first part of Theorem 3.4, we can show that (Exercise 11)

$$\hat{u}_j = h^2 (aw^j + \beta + \gamma\, \overline{w}^j)^{-1} \hat{f}_j, \tag{3.12}$$

provided that $aw^j + \beta + \gamma\overline{w}^j$ is never 0 (see Exercise 12). This gives us $\hat{u}$. We get u by applying the inverse DFT to $\hat{u}$.

Of course, there are also direct methods for finding the u_k [by solving the linear equations in (3.11) for the u_k]. However, often information on the frequency components of the solution is desired. In this case, equation (3.12) provides the discrete Fourier transform of the solution directly.

3.2 Discrete Signals

Sampling invariably produces digitized signals, not continuous ones. The signal becomes a sequence, $x = (\dots, x_{-2}, x_{-1}, x_0, x_1, \dots)$, and the index k in x_k corresponds to the *discrete time*. The graph of a discrete signal is called a *time series*. In this section, we discuss discrete versions of many of the Fourier analysis topics discussed in Chapter 2.

3.2.1 Time Invariant, Discrete Linear Filters

This section should be considered as the discrete analog of Section 2.3.1, where continuous time invariant filters were discussed. We first define the time translation operator T_p on a sequence x by

$$[T_p(x)]_k = x_{k-p}.$$

In words, T_p takes a sequence x and shifts it p units to the right. For example, if $x_k = (1 + k^2)^{-1}$, then $y = T_p(x)$ has entries $y_k = (1 + (k - p)^2)^{-1}$. Note that T_p is a linear operator.

To avoid questions concerning convergence of infinite series, we temporarily assume that our sequences are finite; that is, we consider a sequence x such that $x_k = 0$ for all $|k| > K$, where K is some positive number.

DEFINITION 3.8 *A linear operator* $F : x \rightarrow y$ *that takes a sequence* x *into another sequence* y *is* time invariant *if* $F(T_p(x)) = T_p(F(x))$.

This condition is simply the discretized version of the time invariance condition associated with continuous-time filters (Definition 2.13).

Any linear operator F that takes sequences to sequences is completely determined by what it does to the unit sequences, e^n, where

$$e_k^n = \begin{cases} 0 & k \neq n \\ 1 & k = n. \end{cases}$$

Why? A sequence x can be written as the sum $x = \sum_{n \in Z} x_n e^n$, because its kth component is the sum $\sum_{n \in Z} x_n e_k^n = x_k$. The linearity of F then implies

$$F(x) = \sum_{n \in Z} F(x_n e^n) = \sum_{n \in Z} x_n F(e^n). \qquad (3.13)$$

So knowing the values of F on e^n completely determines F.

Let $f^n = F(e^n)$ be the sequence that e^n is transformed into by F. If F is time invariant, then

$$T_p(f^n) = T_p(F(e^n))$$
$$= F(T_p(e^n)) \quad \text{by time invariance}$$
$$= F(e^{n+p}) \quad \text{(see Exercise 1).}$$

Therefore,

$$T_p(f^n) = f^{n+p}. \qquad (3.14)$$

We also have $(T_p(f^n))_k = f_{k-p}^n$ from the definition of T_p. Therefore, (3.14) becomes $f_k^{n+p} = f_{k-p}^n$. If we set $n = 0$, then $f_k^p = f_{k-p}^0$. Replacing p by n then gives

$$f_k^n = f_{k-n}^0.$$

On the other hand, from (3.13),

$$F(x)_k = \sum_{n \in Z} x_n \underbrace{F(e^n)_k}_{f_k^n}.$$

Since $f_k^n = f_{k-n}^0$, we have

$$F(x)_k = \sum_{n \in Z} x_n f_{k-n}, \quad \text{where } f := f^0.$$

Compare this formula to that in Theorem 2.17 for time invariant operators in the continuous case. There, the operator had the form of a convolution of two continuous functions (see Definition 2.9). If we replace the integral in the continuous convolutions with a sum, we obtain the preceding formula. Let us therefore define the discrete version of the convolution of two sequences.

DEFINITION 3.9 (Discrete Convolution)   *Given the sequences x and y, the convolution $x * y$ is defined to be*

$$(x * y)_k = \sum_{n \in Z} x_{k-n} y_n, \tag{3.15}$$

provided that the series involved are absolutely convergent.

Our analysis in this section can be summarized as follows.

THEOREM 3.10   *If F is a time invariant linear operator acting on sequences, it has the form of a convolution; namely, there is a sequence f such that*

$$F(x) = f * x,$$

provided that the series involved are absolutely convergent. Conversely, if $F(x) = f * x$, then F is a discrete, time invariant linear operator.

We again call such convolution operators *discrete filters*. The sequence f, which satisfies $F(e^0) = f$ and is thus a response to an "impulse" at discrete time 0, is called the *impulse response* (IR). (Refer to the discussion following Theorem 2.17.) If f has an infinite number of nonvanishing terms, it is called an *infinite impulse response* (IIR); if it has only a finite number of nonzero terms, it is a *finite impulse response* (FIR).

3.2.2   Z-Transform and Transfer Functions

In this section, we generalize the discrete Fourier transform to infinite sequences in l^2. The resulting transform is called the *Z-transform*. There is a close relationship between the Z-transform and the complex form of Fourier series. We will encounter the Z-transform again in Chapter 7.

Recall that l^2 is the space of all (complex-valued) sequences having finite energy; that is, all sequences $x = (\ldots, x_{-1}, x_0, x_1, \ldots)$ with $\sum_n |x_n|^2 < \infty$. The inner product of two sequences x and y in l^2 is given by $\langle x, y \rangle = \sum_n x_n \overline{y_n}$.

DEFINITION 3.11   *The Z-transform of a sequence $x = (\ldots, x_{-1}, x_0, x_1, \ldots) \in l^2$ is the function $\widehat{x} : [-\pi, \pi] \to C$:*

$$\widehat{x}(\phi) = \sum_{j=-\infty}^{\infty} x_j e^{-ij\phi}.$$

Often, $z = e^{i\phi}$ is used in the equation defining $\widehat{x}(\cdot)$, so that $\widehat{x}$ becomes

$$\widehat{x}(z) = \sum_{j=-\infty}^{\infty} x_j z^{-j}.$$

This is the origin of the Z in the transform's name.

The Z-transform generalizes the definition of the discrete Fourier transform for finite sequences. To see this, suppose $x = x_0, x_1, \ldots, x_{n-1}$ is a finite sequence (so $x_j = 0$ for $j < 0$ and $j \geq n$). Inserting the angle $\phi = \dfrac{2\pi k}{n}$ into the Z-transform gives

$$\widehat{x}(\frac{2\pi k}{n}) = \sum_{j=0}^{n-1} x_j e^{\frac{-ijk2\pi}{n}} = \sum_{j=0}^{n-1} x_j \overline{w}^{jk}$$

where $w = e^{\frac{2\pi i}{n}}$. The right side is the definition of $\widehat{x}_k$—the kth coefficient of the discrete Fourier transform of x.

Connection with Fourier Series. There is an important relation between the Z-transform and Fourier series. For a given $f \in L^2[-\pi, \pi]$, its Fourier series is

$$f(\phi) = \sum_{n=-\infty}^{\infty} x_n e^{in\phi}$$

where x_n is the nth Fourier coefficient of f:

$$x_n = \frac{1}{2\pi} \int_{-\pi}^{\pi} f(\phi) e^{-in\phi} \, d\phi.$$

From the definition of $\widehat{x}$,

$$\widehat{x}(\phi) = \sum_{n=-\infty}^{\infty} x_n e^{-in\phi} = f(-\phi). \tag{3.16}$$

Thus, a Fourier series expansion is a process that converts a function $f \in L^2[-\pi, \pi]$ into a sequence, $\{x_n\} \in l^2$; and the Z-transform takes the sequence $\{x_n\}$ back to the function f. The following diagram illustrates this relationship.

$$f \in L^2[-\pi, \pi] \qquad \xrightarrow{\quad\text{Fourier Series}\quad} \qquad \{x_n\} \in l^2$$

$$f(-\phi) = \widehat{x}(\phi) \in L^2[-\pi, \pi] \quad \xleftarrow{\quad\text{Z-Transform}\quad} \quad \{x_n\} \in l^2$$

Isometry between L^2 and l^2. A further connection to Fourier series is given by Parseval's theorem [see equation (1.39)]. If x_n and y_n are the Fourier coefficients of f and $g \in L^2[-\pi, \pi]$, then

$$\sum_{n=-\infty}^{\infty} x_n \overline{y_n} = \frac{1}{2\pi} \int_{-\pi}^{\pi} f(\phi)\overline{g(\phi)} \, d\phi \quad \text{from (1.39)}$$

$$= \frac{1}{2\pi} \int_{-\pi}^{\pi} \widehat{x}(-\phi)\overline{\widehat{y}(-\phi)} \, d\phi \quad \text{from (3.16).}$$

With the change of variables $\phi \to -\phi$, we obtain

$$\frac{1}{2\pi} \int_{-\pi}^{\pi} \widehat{x}(\phi)\overline{\widehat{y}(\phi)} \, d\phi = \sum_{n=-\infty}^{\infty} x_n \overline{y_n}$$

or, in other words,

$$\frac{1}{2\pi} \langle \widehat{x}, \widehat{y} \rangle_{L^2[-\pi,\pi]} = \langle x, y \rangle_{l^2}$$

where $x = (\ldots, x_{-1}, x_0, x_1, \ldots)$ and $y = (\ldots, y_{-1}, y_0, y_1, \ldots)$. We summarize this discussion in the following theorem.

THEOREM 3.12 *The Z-transform is an isometry between l^2 and $L^2[-\pi, \pi]$ (preserves the inner products up to a scalar factor of $1/2\pi$). In particular, if $x = (\ldots, x_{-1}, x_0, x_1, \ldots)$ and $y = (\ldots, y_{-1}, y_0, y_1, \ldots)$ are sequences in l^2, then*

$$\frac{1}{2\pi} \langle \widehat{x}, \widehat{y} \rangle_{L^2[-\pi,\pi]} = \langle x, y \rangle_{l^2}.$$ ♦

Convolutions. Now we discuss the relationship between convolution operators and the Z-transform. We have already mentioned that convolution operators resemble their continuous counterparts. The following theorem carries this analogy one step further (compare with Theorem 2.10).

THEOREM 3.13 *Suppose $f = \{f_n\}$ and $x = \{x_n\}$ are sequences in l^2. Then*

$$\widehat{(f * x)}(\phi) = \widehat{f}(\phi)\widehat{x}(\phi).$$

The function $\widehat{f}(\phi)$ is the Z-transform of the sequence f and is also called the *transfer function* associated to the operator F, where $F(x) = f * x$. Often, we identify the convolution operator F with its associated sequence f, and therefore we also identify their transfer functions (i.e., $\widehat{F}$ with $\widehat{f}$).

Proof Using the definition of the Z-transform , we have

$$\widehat{(f * x)}(\phi) = \sum_{n=-\infty}^{\infty} (f * x)(n) e^{-in\phi} = \sum_{n=-\infty}^{\infty} \left(\sum_{k=-\infty}^{\infty} f_k x_{n-k} \right) e^{-in\phi}.$$

Writing $e^{-in\phi} = e^{-ik\phi} e^{-i(n-k)\phi}$, we obtain

$$\widehat{(f * x)}(\phi) = \sum_{n=-\infty}^{\infty} \sum_{k=-\infty}^{\infty} f_k e^{-ik\phi} x_{n-k} e^{-i(n-k)\phi}$$

$$= \sum_{k=-\infty}^{\infty} f_k e^{-ik\phi} \sum_{n=-\infty}^{\infty} x_{n-k} e^{-i(n-k)\phi}$$

where we have switched the order of summation. By replacing $n - k$ by the index m in the second sum, this equation becomes

$$\widehat{(f * x)}(\phi) = \left(\sum_{k=-\infty}^{\infty} f_k e^{-ik\phi} \right) \left(\sum_{m=-\infty}^{\infty} x_m e^{-im\phi} \right) = \widehat{f}(\phi)\widehat{x}(\phi)$$

as desired.   ◆

Adjoint of Convolution Operators. Recall that the adjoint of an operator $F : l^2 \to l^2$ is the operator $F^* : l^2 \to l^2$, which is defined by (see Definition 0.29):

$$\langle F(x), y \rangle = \langle x, F^*(y) \rangle \quad \text{for } x, y \in l^2.$$

The next theorem calculates a formula for the sequence associated to the adjoint of a convolution operator. This theorem will be needed in Chapter 7.

THEOREM 3.14 *Suppose F is the convolution operator associated with the sequence f_n. Then F^* is the convolution operator associated to the sequence $f_n^* = \overline{f}_{-n}$. The transfer function for F^* is $\overline{\widehat{F}(\phi)}$.*

Proof From the definition of convolution and l^2, we have

$$\langle F(x), y \rangle_{l^2} = \langle f * x, y \rangle_{l^2}$$

$$= \sum_{n=-\infty}^{\infty} (f * x)_n \overline{y}_n$$

$$= \sum_{n=-\infty}^{\infty} \sum_{k=-\infty}^{\infty} f_{n-k} x_k \overline{y}_n$$

$$= \sum_{k=-\infty}^{\infty} x_k \overline{\sum_{n=-\infty}^{\infty} \overline{f}_{-(k-n)} y_n}.$$

The last sum on the right (under the big conjugation) is $(f^* * y)(k)$, where $f^*(m) = \overline{f}_{-m}$. Therefore, we obtain

$$\langle F(x), y \rangle_{l^2} = \sum_{k=-\infty}^{\infty} x_k \overline{(f^* * y)(k)} = \langle x, (f^* * y) \rangle_{l^2}.$$

Since $\langle F(x), y \rangle = \langle x, F^*(y) \rangle$, by definition, we conclude that $F^*(y) = (f^* * y)$, where $f_m^* = \overline{f}_{-m}$, as desired.

The second part of this theorem follows from the first because

$$\widehat{F^*}(\phi) = \sum_{n=-\infty}^{\infty} f_n^* e^{-in\phi} \quad \text{(by definition of } \widehat{\;})$$

$$= \sum_{n=-\infty}^{\infty} \overline{f}_{-n} e^{-in\phi} \quad \text{(by the first part of the theorem)}$$

$$= \overline{\sum_{n=-\infty}^{\infty} f_{-n} e^{in\phi}}$$

$$= \overline{\sum_{m=-\infty}^{\infty} f_m e^{-im\phi}} \quad \text{(by letting } m = -n)$$

$$= \overline{\widehat{F}}(\phi)$$

as desired. ◆

3.3 Exercises

Some of the following problems require the fast Fourier transform (e.g., Maple or MATLAB's FFT routine).

1. Show that $T_p(e^n) = e^{n+p}$, where T_p and e^n are defined in Section 3.2.

2. Prove Theorem 3.4.

3. Derive equation (3.8) (see [3, §6.4]).

4. Plot the solution u in (3.8), given $a = 1$, $b = 2$, $c = 37$, and $f_0 = 1$.

5. In the previous exercise, use MATLAB or some similar program to take 256 samples of u on the interval $[0, 4]$. Plot the absolute value of the FFT of u. Locate the largest natural frequency $\omega/2\pi$. Compare with the result from the previous exercise.

6. The FFT locates the frequency reasonably well. Find a way to estimate the damping constant, μ, from FFT data.

7. *Filtering.* Let

$$f(t) = e^{-t^2/10} \left(\sin(2t) + 2\cos(4t) + 0.4\sin(t)\sin(50t) \right).$$

Discretize f by setting $y_k = f(2k\pi/256)$, $k = 1, \ldots, 256$. Use the fast Fourier transform to compute $\widehat{y}_k$ for $0 \le k \le 256$. Recall from Theorem 3.4 that $y_{n-k} = \overline{y}_k$. Thus, the low-frequency coefficients are $\widehat{y}_0, \ldots, \widehat{y}_m$ and

$\widehat{y}_{256-m}, \ldots, \widehat{y}_{256}$ for some low value of m. Filter out the high-frequency terms by setting $\widehat{y}_k = 0$ for $m \le k \le 255 - m$ with $m = 6$; then apply the inverse fast Fourier transform to this new set of $\widehat{y}_k$ to compute the y_k (now filtered); plot the new values of y_k and compare with the original function. Experiment with other values of m.

8. *Compression.* Let $tol = 1.0$. In Exercise 7, if $|\widehat{y}_k| < tol$ then set $\widehat{y}_k$ equal to zero. Apply the inverse fast Fourier transform to this new set of $\widehat{y}_k$ to compute the y_k; plot the new values of y_k and compare with the original function. Experiment with other values of tol. Keep track of the percentage of Fourier coefficients that have been filtered out. MATLAB's sort command is useful for finding a value for tol in order to filter out a specified percentage of coefficients (see the compression routine in Appendix B on MATLAB code). Compute the relative l^2-error of the compressed signal as compared with the original signal (again, see Appendix B on MATLAB code).

9. Repeat the previous two exercises over the interval $0 \le x \le 1$ with the function

$$f(x) = -52x^4 + 100x^3 - 49x^2 + 2 + N(100(x - 1/3)) + N(200(x - 2/3))$$

where

$$N(t) = te^{-t^2}.$$

This time, $y_k = f(k/256)$.

10. Derive equation (3.11) by inserting (3.9) and (3.10) into (3.7) at $t = t_k = 2\pi k/n$.

11. Derive equation (3.12), assuming that $aw^j + \beta + \gamma \overline{w}^j$ is never 0.

12. Let a, b, c, and $4ac - b^2$ be positive. As in (3.11), set $\beta = ch^2 + bh - 2a$ and $\gamma = a - bh$. Show that if $bh < 2a$, $aw^j + \beta + \gamma \overline{w}^j$ is never 0. (*Hint:* Show that the quadratic equation $az^2 + \beta z + \gamma = 0$ has no solutions for which $|z| = 1$.)

13. Find the exact, $\pi/3$-periodic solution to $u'' + 2u' + 2u = 3\cos(6t)$. Compare it with the discrete approximation obtained for these values of n: $n = 16$, $n = 64$, and $n = 256$. You will need to use MATLAB or similar software to compute the FFT and inverse FFT involved.

14. Recall that the complex exponentials $u = \exp(int)$ are 2π-periodic eigenfunctions of the operator $D^2[u] = u''$ (this means that $D^2[u] = -n^2 u$). A discretized version of this operator acts on the periodic sequences in $\mathcal{S}_n$. If u_k is n-periodic, then define $L[u]$ via

$$L[u]_k = u_{k+1} + u_{k-1} - 2u_k.$$

(a) Show that L maps $\mathcal{S}_n$ into itself.

(b) For $n = 4$, show the the matrix M_4 that represents L is

$$M_4 = \begin{pmatrix} -2 & 1 & 0 & 1 \\ 1 & -2 & 1 & 0 \\ 0 & 1 & -2 & 1 \\ 1 & 0 & 1 & -2 \end{pmatrix}.$$

(c) Observe that M_4 is self-adjoint and thus diagonalizable. Find its eigenvalues and corresponding eigenvectors. How are these related to the matrix columns for the matrix F_4 in (3.2)? Could you have diagonalized this matrix via an FFT? Explain.

(d) Generalize this result for all n. (*Hint*: Use the DFT on $L[u]_k$—recall that the FFT is really a fast DFT.)

15. (Circulants and the DFT.) An $n \times n$ matrix A is called a *circulant* if all of its diagonals (main, super, and sub) are constant and the indices are interpreted "modulo n." For example, this 4×4 matrix is a circulant:

$$\begin{pmatrix} 9 & 2 & 1 & 7 \\ 7 & 9 & 2 & 1 \\ 1 & 7 & 9 & 2 \\ 2 & 1 & 7 & 9 \end{pmatrix}.$$

(a) Look at the n-periodic sequence a, where $a_\ell = A_{\ell+1,1}$, $\ell = 0 \ldots n-1$. Write the entries of A in terms of the sequence a.

(b) Let X be an $n \times 1$ column vector. Show that $Y = AX$ is equivalent to $y = a * x$ if x, y are n periodic sequences for which $x_\ell = X_{\ell+1,1}$ and similarly for $y_\ell = Y_{\ell+1,1}$, $\ell = 0, \ldots, n-1$..

(c) Prove that the DFT diagonalizes all circulant matirices; that is, that $n^{-1}F_n^T A\bar{F}_n = D$, where D is diagonal. What are the diagonal entries of D? (*i.e.*, what are the eigenvalues of A)?

16. Find the Z-transform for the sequence

$$x = \begin{pmatrix} \cdots & 0 & 0 & 1 & \frac{1}{2} & \frac{1}{4} & \cdots & \frac{1}{2^n} & \cdots \end{pmatrix}.$$

Chapter 4

Haar Wavelet Analysis

4.1 Why Wavelets?

Wavelets were first applied in geophysics to analyze data from seismic surveys, which are used in oil and mineral exploration to get "pictures" of layering in subsurface rock. In fact, geophysicists rediscovered them; mathematicians had developed them to solve abstract problems some twenty years earlier but had not anticipated their applications in signal processing.[1]

Seismic surveys are made up of many two-dimensional pictures or slices, which are sewn together to give a three-dimensional image of the structure of rock below the surface. Each slice is obtained by placing geophones—seismic "microphones"—at equally spaced intervals along a line, the seismic line. Dynamite is set off at one end of the line to create a seismic wave in the ground. Every geophone along the line records the movement of the earth due to the blast, from start to finish; this record is its *seismic trace* (see Figure 1).

The first wave recorded by the geophones is the direct wave, which travels along the surface. This is usually not important. Subsequent waves are reflected off rock layers below ground. These are the important ones. Knowledge of the time that the wave hits a geophone gives information about where the layer that reflected it is located. The "wiggles" that the wave produces tell something about the fine details of the layer. The traces from the all the geophones on a line can be combined to give the slice for the ground directly beneath the line.

The key to an accurate seismic survey is a proper analysis of each trace. The Fourier transform is not a good tool here. It can only provide frequency information (the oscillations that comprise the signal). It gives no direct in-

---

[1]See Meyer's book [13] for an interesting, first-hand account of how wavelets developed.

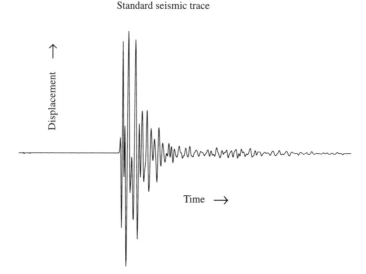

Figure 1 A typical seismic trace. Displacement is plotted versus time. Both the oscillations and the time they occur are important.

formation about *when* an oscillation occurred. Another tool, the short-time Fourier transform, is better. The full-time interval is divided into a number of small, equal-time intervals; these are individually analyzed using the Fourier transform. The result contains time and frequency information. However, there is a problem with this approach. The equal-time intervals are not adjustable; the times when very-short-duration, high-frequency bursts occur are hard to detect.

Enter wavelets. Wavelets can keep track of time and frequency information. They can be used to "zoom in" on the short bursts mentioned previously, or to "zoom out" to detect long, slow oscillations.

4.2   Haar Wavelets

4.2.1   The Haar Scaling Function

There are two functions that play a primary role in wavelet analysis, the scaling function ϕ and the wavelet ψ. These two functions generate a family of functions that can be used to break up or reconstruct a signal. To emphasize the "marriage" involved in building this "family," ϕ is sometimes called the "father wavelet" and ψ, the "mother wavelet."

The simplest wavelet analysis is based on the *Haar scaling function* , whose graph is given in Figure 4. The building blocks are translations and dilations (both in height and width) of this basic graph.

We want to illustrate the basic ideas involved in such an analysis. Consider the signal shown in Figure 2. We may think of this as a measurement of some physical quantity—perhaps line voltage over a single cycle—as a function of time. The two sharp spikes in the graph might represent noise coming from a loose connection in the volt meter, and we want to filter out this undesirable noise. The graph in Figure 3 shows one possible approximation to the signal using Haar building blocks. The high-frequency noise shows up as tall, thin blocks. An algorithm that deletes the thin blocks will eliminate the noise and not disturb the rest of the signal.

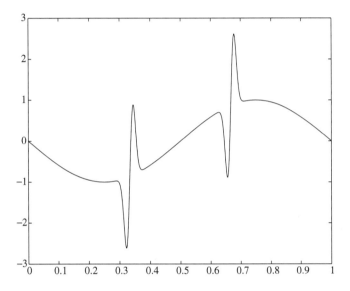

Figure 2 Voltage from a faulty meter

The building blocks generated by the Haar scaling function are particularly simple and illustrate the general ideas underlying a multiresolution analysis, which we will discuss in detail. The disadvantage of the Haar wavelets is that they are discontinuous and therefore do not approximate continuous signals very well (Figure 3 does not really approximate Figure 2 too well). In later sections, we introduce other wavelets, due to Daubechies, that are continuous but still retain the localized behavior exhibited by the Haar wavelets.

4.2.2   Basic Properties of the Haar Scaling Function

DEFINITION 4.1   *The Haar scaling function is defined as*

$$\phi(x) = \begin{cases} 1, & \text{if } 0 \leq x < 1 \\ 0, & \text{elsewhere.} \end{cases}$$

The graph of the Haar scaling function is given in Figure 4.

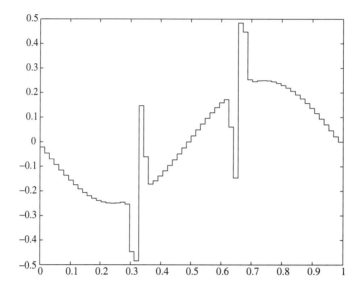

Figure 3 Approximation of voltage signal by Haar functions

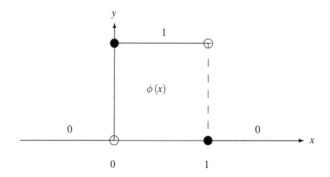

Figure 4 Graph of the Haar scaling function

The function $\phi(x - k)$ has the same graph as ϕ but translated to the right by k units (assuming k is positive). Let V_0 be the space of all functions of the form

$$\sum_{k \in Z} a_k \phi(x - k) \quad a_k \in R$$

where k can range over any finite set of positive or negative integers. Since $\phi(x - k)$ is discontinuous at $x = k$ and $x = k + 1$, an alternative description of V_0 is that it consists of all piecewise constant functions whose discontinuities are contained in the set of integers. Since k ranges over a finite set, each element of V_0 is zero outside a bounded set. Such a function is said to have *finite or compact support*. The graph of a typical element of V_0 is given in Figure 5.

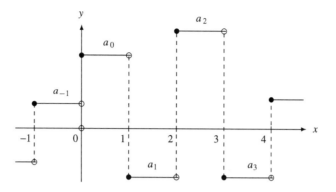

Figure 5 Graph of typical element in V_0

Note that a function in V_0 may not have discontinuities at all the integers (for example, if $a_1 = a_2$, then the preceding sum is continuous at $x = 2$).

EXAMPLE 4.2

The graph of the function

$$f(x) = 2\phi(x) + 3\phi(x-1) + 3\phi(x-2) - \phi(x-3) \in V_0$$

is given in Figure 6. This function has discontinuities at $x = 0, 1, 3$, and 4. ∎

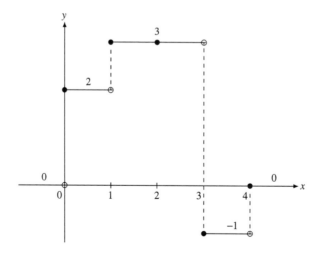

Figure 6 Plot of f in Example 4.2

We need blocks that are thinner to analyze signals of high frequency. The building block whose width is half that of the graph of ϕ is given by the graph of $\phi(2x)$ shown in Figure 7.

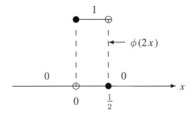

Figure 7 Graph of $\phi(2x)$

The function $\phi(2x - k) = \phi(2(x - k/2))$ is the same as the graph of the function of $\phi(2x)$ but shifted to the right by $k/2$ units. Let V_1 be the space of functions of the form

$$\sum_{k \in Z} a_k \phi(2x - k) \quad a_k \in R.$$

Geometrically, V_1 is the space of piecewise constant functions of finite support with possible discontinuities at the half-integers $\{0, \pm 1/2, \pm 1, \pm 3/2, \dots\}$.

EXAMPLE 4.3

The graph of the function

$$f(x) = 4\phi(2x) + 2\phi(2x - 1) + 2\phi(2x - 2) - \phi(2x - 3) \in V_0$$

is given in Figure 8. This function has discontinuities at $x = 0, 1/2, 3/2,$ and 2. ∎

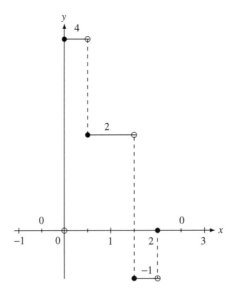

Figure 8 Plot of f in Example 4.3

We make the following more general definition.

DEFINITION 4.4 *Suppose j is any nonnegative integer. The space of step functions at level j, denoted by V_j, is defined to be be the space spanned by the set*

$$\{\ldots, \phi(2^j x + 1), \phi(2^j x), \phi(2^j x - 1), \phi(2^j x - 2), \ldots\}$$

over the real numbers. V_j is the space of piecewise constant functions of finite support whose discontinuities are contained in the set

$$\{\ldots, -1/2^j, 0, 1/2^j, 2/2^j, 3/2^j, \ldots\}.$$

A function in V_0 is a piecewise constant function with discontinuities contained in the set of integers. Any function in V_0 is also contained in V_1, which consists of piecewise constant functions whose discontinuities are contained in the set of half-integers $\{\ldots, -1/2, 0, 1/2, 1, 3/2, \ldots\}$. The same applies for $V_1 \subset V_2$ and so forth:

$$V_0 \subset V_1 \subset \cdots V_{j-1} \subset V_j \subset V_{j+1} \cdots.$$

This containment is strict. For example, the function $\phi(2x)$ belongs to V_1 but does not belong to V_0 [since $\phi(2x)$ is discontinuous at $x = 1/2$].

V_j contains all relevant information up to a resolution scale of order 2^{-j}. As j gets larger, the resolution gets finer. The fact that $V_j \subset V_{j+1}$ means that no information is lost as the resolution gets finer. This containment relation is also the reason why V_j is defined in terms of $\phi(2^j x)$ instead of $\phi(ax)$ for some other factor a. If, for example, we had defined V_2 using $\phi(3x - j)$ instead of $\phi(4x - j)$, then V_2 would not contain V_1 (since the set of multiples of $1/2$ is not contained in the set of multiples of $1/3$).

4.2.3 Basic Properties of the Haar Scaling Function

The following theorem is an easy consequence of the definitions.

THEOREM 4.5

- *A function $f(x)$ belongs to V_0 if and only if $f(2^j x)$ belongs to V_j.*

- *A function $f(x)$ belongs to V_j if and only if $f(2^{-j} x)$ belongs to V_0.*

Proof To prove the first statement, if a function f belongs to V_0, then $f(x)$ is a linear combination of $\{\phi(x - k), k \in Z\}$. Therefore, $f(2^j x)$ is a linear combination of $\{\phi(2^j x - k), k \in Z\}$, which means that $f(2^j x)$ is a member of V_j. The proofs of the converse and the second statement are similar.   ♦

The graph of the function $\phi(2^j x)$ is a spike of width $1/2^j$. When j is large, the graph of $\phi(2^j x)$ (appropriately translated) is similar to one of the spikes of a signal that we may wish to filter out. Thus it is desirable to have an efficient algorithm to decompose a signal into its V_j-components. One way to perform this decomposition efficiently is to construct an orthonormal basis for V_j (using the L^2 inner product).

Let's start with V_0. This space is generated by ϕ and its translates. The functions $\phi(x - k)$ each have unit norm in L^2; that is,

$$\|\phi(x - k)\|_{L^2}^2 = \int_{-\infty}^{\infty} \phi(x - k)^2 \, dx = \int_k^{k+1} 1 \, dx = 1.$$

If j is different from k, then $\phi(x - j)$ and $\phi(x - k)$ have disjoint supports (see Figure 9). Therefore,

$$\langle \phi(x - j), \, \phi(x - k) \rangle_{L^2} = \int_{-\infty}^{\infty} \phi(x - j)\phi(x - k) \, dx = 0 \quad j \neq k$$

and so the set $\{\phi(x - k), k \in Z\}$ is an orthonormal basis for V_0.

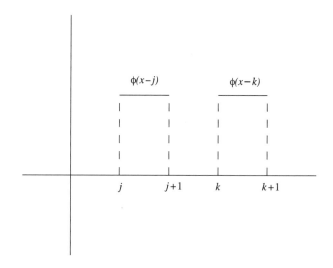

Figure 9 $\phi(x - j)$ and $\phi(x - k)$ have disjoint support

The same argument establishes the following more general result.

THEOREM 4.6   *The set of functions $\{2^{j/2}\phi(2^j x - k); \ k \in Z\}$ is an orthonormal basis of V_j.*

[The factor $2^{j/2}$ is present since $\int_{-\infty}^{\infty} (\phi(2^j x))^2 \, dx = 1/2^j$.]

4.2.4   The Haar Wavelet

Having an orthonormal basis of V_j is only half of the picture. In order to solve our noise-filtering problem, we need to have a way of isolating the "spikes" that belong to V_j but that are not members of V_{j-1}. This is where the wavelet ψ enters the picture.

The idea is to decompose V_j as an orthogonal sum of V_{j-1} and its complement. Again, let's start with $j = 1$ and identify the orthogonal complement of V_0 in V_1. Since V_0 is generated by ϕ and its translates, it is reasonable to expect that the orthogonal complement of V_0 is generated by the translates of some function ψ. Two key facts are needed to construct ψ:

1. ψ is a member of V_1 and so ψ can be expressed as $\psi(x) = \sum_l a_l \phi(2x - l)$ for some choice of $a_l \in R$ (and only a finite number of the a_l are nonzero).

2. ψ is orthogonal to V_0. This is equivalent to $\int \psi(x)\phi(x - k)\,dx = 0$ for all integers k.

The first requirement means that ψ is built from blocks of width $1/2$ (i.e., scalar multiples of Figure 7 and its translates). The second requirement with $k = 0$ implies $\int_{-\infty}^{\infty} \psi(x)\phi(x)\,dx = 0$. The simplest ψ satisfying both of these requirements is the function whose graph appears in Figure 10. This graph consists of two blocks of width one-half and can be written as

$$\psi(x) = \phi(2x) - \phi(2(x - 1/2)) = \phi(2x) - \phi(2x - 1)$$

thus satisfying the first requirement. In addition,

$$\int_{-\infty}^{\infty} \phi(x)\psi(x)\,dx = \int_0^{1/2} 1\,dx - \int_{1/2}^1 1\,dx$$
$$= 1/2 - 1/2$$
$$= 0.$$

Thus, ψ is orthogonal to ϕ. If $k \neq 0$, then the support of $\psi(x)$ and the support of $\phi(x - k)$ do not overlap and so $\int \psi(x)\phi(x - k)\,dx = 0$. Therefore, ψ belongs to V_1 and is orthogonal to V_0; ψ is called the *Haar wavelet*.

DEFINITION 4.7   *The Haar wavelet is the function*

$$\psi(x) = \phi(2x) - \phi(2x - 1).$$

Its graph is given in Figure 10.

You can show (see Exercise 4) that any function

$$f_1 = \sum_k a_k \phi(2x - k) \in V_1$$

is orthogonal to V_0—that is, orthogonal to each $\phi(x - l), l \in Z$) if and only if

$$a_1 = -a_0, \quad a_3 = -a_2, \ldots.$$

In this case,

$$f_1 = \sum_{k \in Z} a_{2k} \left(\phi(2x - 2k) - \phi(2x - 2k - 1) \right) = \sum_{k \in Z} a_{2k} \psi(x - k). \qquad (4.1)$$

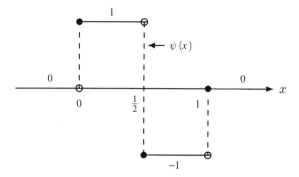

Figure 10 The Haar wavelet $\psi(x)$

In other words, a function in V_1 is orthogonal to V_0 if and only if it is of the form $\sum_k a_k \psi(x-k)$ (relabeling a_{2k} by a_k).

Let W_0 be the space of all functions of the form

$$\sum_{k \in Z} a_k \psi(x-k) \quad a_k \in R$$

where, again, we assume that only a finite number of the a_k are nonzero. What we have just shown is that W_0 is the orthogonal complement of V_0 in V_1 or, in other words, $V_1 = V_0 \oplus W_0$ (recall from Chapter 0 that $\oplus$ means that V_0 and W_0 are orthogonal to each other).

In a similar manner, the following more general result can be established.

THEOREM 4.8 *Let W_j be the space of functions of the form*

$$\sum_{k \in Z} a_k \psi(2^j x - k) \quad a_k \in R$$

where we assume that only a finite number of a_k are nonzero. W_j is the orthogonal complement of V_j in V_{j+1} and

$$V_{j+1} = V_j \oplus W_j.$$

Proof To establish this theorem, we must show two facts:

1. Each function in W_j is orthogonal to every function in V_j.

2. Any function in V_{j+1} that is orthogonal to V_j must belong to W_j.

For the first requirement, suppose that $g = \sum_{k \in Z} a_k \psi(2^j x - k)$ belongs to W_j and suppose that f belongs to V_j. We must show

$$\langle g, f \rangle_{L^2} = \int_{-\infty}^{\infty} g(x)\overline{f(x)}\, dx = 0.$$

Since $f(x)$ belongs to V_j, the function $f(2^{-j}x)$ belongs to V_0. So

$$0 = \int_{-\infty}^{\infty} \sum_{k \in Z} a_k \psi(x-k)\overline{f(2^{-j}x)}\, dx \quad \text{(because } \psi \text{ is orthogonal to } V_0\text{)}$$

$$= 2^j \int_{-\infty}^{\infty} \sum_{k \in Z} a_k \psi(2^j y - k)\overline{f(y)}\, dy \quad \text{(by letting } y = 2^{-j}x\text{)}$$

$$= 2^j \int_{-\infty}^{\infty} g(y)\overline{f(y)}\, dy.$$

Therefore, g is orthogonal to any $f \in V_j$ and the first requirement has been established.

The discussion leading to equation (4.1) establishes the second requirement when $j = 0$ since we showed that any function in V_1 that is orthogonal to V_0 must be a linear combination of $\{\psi(x-k),\ k \in Z\}$. The argument for general j is very similar to the case when $j = 0$. ♦

By successively decomposing V_j, V_{j-1} and so on, we have

$$\begin{aligned}
V_j &= W_{j-1} \oplus V_{j-1} \\
&= W_{j-1} \oplus W_{j-2} \oplus V_{j-2} \\
&\quad \cdots \\
&= W_{j-1} \oplus W_{j-2} \oplus \cdots \oplus W_0 \oplus V_0.
\end{aligned}$$

So each f in V_j can be decomposed uniquely as a sum

$$f = w_{j-1} + w_{j-2} + \cdots + w_0 + f_0$$

where each w_l belongs to W_l, $0 \le l \le j-1$ and f_0 belongs to V_0. Intuitively, w_l represents the "spikes" of f of width $1/2^{l+1}$ that cannot be represented as linear combinations of spikes of other widths.

What happens when j goes to infinity? The answer is provided in the following theorem.

THEOREM 4.9 *The space $L^2(R)$ can be decomposed as an infinite orthogonal direct sum*

$$L^2(R) = V_0 \oplus W_0 \oplus W_1 \oplus \cdots .$$

In particular, each $f \in L^2(R)$ can be written uniquely as

$$f = f_0 + \sum_{j=0}^{\infty} w_j$$

where f_0 belongs to V_0 and w_j belongs to W_j.

The infinite sum should be thought of as a limit of finite sums. In other words,

$$f = f_0 + \lim_{N \to \infty} \sum_{j=0}^{N} w_j \tag{4.2}$$

where the limit is taken in the sense of L^2. Although the proof of this result is beyond the scope of this text, some intuition can be given. There are two key ideas. The first is that any function in $L^2(R)$ can be approximated by continuous functions (for an intuitive explanation of this idea, see Lemma 1.38). The second is that any continuous function can be approximated as closely as desired by a step function whose discontinuities are multiples of 2^{-j} for suitably large j (see Figure 11). Such a step function, by definition, belongs to V_j. The theorem is then established by putting both ideas together.

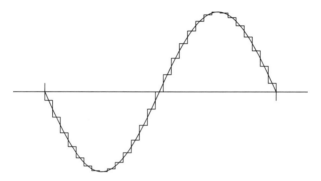

Figure 11 Approximating by step functions

4.3 Haar Decomposition and Reconstruction Algorithms

4.3.1 Decomposition

Now that V_j has been decomposed as a direct sum of V_0 and W_l for $0 \le l < j$, the solution to our noise-filtering problem is theoretically easy. First, we approximate f by a step function $f_j \in V_j$ (for j suitably large) using Theorem 4.9. Then we decompose f_j into its components

$$f_j = f_0 + w_1 + \cdots + w_{j-1}, \quad w_l \in W_l.$$

The component w_l represents the spikes of width $1/2^{l+1}$. For l sufficiently large, these spikes are thin enough to represent noise. For example, suppose that spikes of width less than 0.01 represent noise. Since $2^{-6} < 0.01 < 2^{-7}$, any w_j with $j \ge 6$ represents noise. To filter out this noise, these components can be set equal to zero. The rest of the sum represents a signal that is still relatively close to f and that is noise free.

In order to implement this theoretical algorithm, an efficient way of performing the decomposition given in Theorem 4.9 is needed. The first step is to approximate the original signal f by a step function of the form

$$f_j(x) = \sum_{l \in Z} a_l \phi(2^j x - l). \tag{4.3}$$

The procedure is to sample the signal at $x = \ldots, -1/2^j, 0, 1/2^j, \ldots$, which leads to $a_l = f(l/2^j)$ for $l \in Z$. An illustration of this procedure is given in Figure 11, where f is the continuous signal and f_j is the step function. Here, j is chosen so that the mesh size 2^{-j} is small enough so that $f_j(x)$ captures the essential features of the signal. The range of l depends on the domain of the signal. If the signal is defined on $0 \leq x \leq 1$, then the range of l is $0 \leq l \leq 2^j - 1$. In general, we will not specify the range of l unless a specific example is discussed.

Now the task is to decompose $\phi(2^j x - l)$ into its W_l-components for $l < j$. The following relations between ϕ and ψ are needed:

$$\phi(2x) = (\psi(x) + \phi(x))/2 \tag{4.4}$$
$$\phi(2x - 1) = (\phi(x) - \psi(x))/2, \tag{4.5}$$

which follow easily by looking at their graphs (see Figures 4 and 10). More generally, we have the following lemma.

LEMMA 4.10 *The following relations hold for all $x \in R$:*

$$\phi(2^j x) = (\psi(2^{j-1}x) + \phi(2^{j-1}x))/2 \tag{4.6}$$
$$\phi(2^j x - 1) = (\phi(2^{j-1}x) - \psi(2^{j-1}x))/2. \tag{4.7}$$

This lemma follows by replacing x by $2^{j-1}x$ in equations (4.4) and (4.5).

This lemma can be used to decompose $\phi(2^j x - l)$ into its W_l-components for $l < j$. An example helps illustrate the process.

EXAMPLE 4.11

Let f be given by the graph in Figure 12. Notice that the mesh size needed to capture all the features of f is 2^{-2}. A description of f in terms of $\phi(2^2 x - l)$ is given by

$$f(x) = 2\,\phi(4x) + 2\phi(4x - 1) + \phi(4x - 2) - \phi(4x - 3). \tag{4.8}$$

We wish to decompose f into its W_1, W_0, and V_0 components. The following equations follow from (4.6) and (4.7) with $j = 2$:

$$\phi(4x) = (\psi(2x) + \phi(2x))/2$$
$$\phi(4x - 1) = (\phi(2x) - \psi(2x))/2$$
$$\phi(4x - 2) = \phi(4(x - 1/2)) = (\psi(2(x - 1/2)) + \phi(2(x - 1/2)))/2$$
$$\phi(4x - 3) = \phi(4(x - 1/2) - 1) = (\phi(2(x - 1/2)) - \psi(2(x - 1/2)))/2.$$

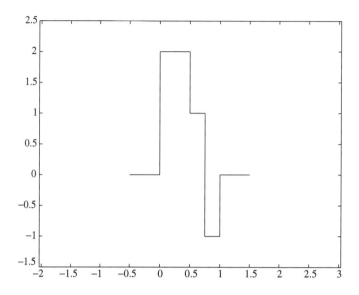

Figure 12 Figure for Example 4.11

Using these equations together with (4.8) and collecting terms yields

$$f(x) = [\psi(2x) + \phi(2x)] + [\phi(2x) - \psi(2x)]$$
$$+ [\psi(2x - 1) + \phi(2x - 1)]/2 - [\phi(2x - 1) - \psi(2x - 1)]/2$$
$$= \psi(2x - 1) + 2\phi(2x).$$

The W_1-component of $f(x)$ is $\psi(2x - 1)$, since W_1 is the linear span of $\{\psi(2x - k); \ k \in Z\}$. The V_1-component of $f(x)$ is given by $2\phi(2x)$. This component can be further decomposed into a V_0-component and a W_0-component by using the equation $\phi(2x) = (\phi(x) + \psi(x))/2$. The result is

$$f(x) = \psi(2x - 1) + \psi(x) + \phi(x).$$

This equation can also be verified geometrically by examining the graphs of the functions involved. The terms in the expression at right are the components of f in W_1, W_0, and V_0, respectively.   ■

Using this example as a guide, we can proceed with the general decomposition scheme as follows. First divide the sum $f_j(x) = \sum_k a_k\phi(2^j x - k)$ into even and odd terms:

$$f_j(x) = \sum_{k \in Z} a_{2k}\phi(2^j x - 2k) + \sum_{k \in Z} a_{2k+1}\phi(2^j x - 2k - 1). \qquad (4.9)$$

Next, we use (4.6) and (4.7) with x replaced by $x - k2^{1-j}$:

$$\phi(2^j x - 2k) = (\psi(2^{j-1}x - k) + \phi(2^{j-1}x - k))/2 \qquad (4.10)$$

$$\phi(2^j x - 2k - 1) = (\phi(2^{j-1}x - k) - \psi(2^{j-1}x - k))/2. \qquad (4.11)$$

Substituting these expressions into (4.9) yields

$$f_j(x) = \sum_{k \in Z} a_{2k} \left(\psi(2^{j-1}x - k) + \phi(2^{j-1}x - k) \right) / 2$$

$$+ \sum_{k \in Z} a_{2k+1} \left(\phi(2^{j-1}x - k) - \psi(2^{j-1}x - k) \right) / 2$$

$$= \sum_{k \in Z} \left(\frac{a_{2k} - a_{2k+1}}{2} \right) \psi(2^{j-1}x - k) + \left(\frac{a_{2k} + a_{2k+1}}{2} \right) \phi(2^{j-1}x - k)$$

$$= w_{j-1} + f_{j-1}.$$

The first term on the right, w_{j-1}, represents the W_{j-1}-component of f_j since W_{j-1} is, by definition, the linear span of $\{\psi(2^{j-1}x - k), \; k \in Z\}$. Likewise, the second term on the right, f_{j-1}, represents the V_{j-1}-component of f_j. We summarize the preceding decomposition algorithm in the following theorem.

THEOREM 4.12 (Haar Decomposition) *Suppose*

$$f_j(x) = \sum_{k \in Z} a_k^j \phi(2^j x - k) \; \in V_j.$$

Then f_j can be decomposed as

$$f_j = w_{j-1} + f_{j-1}$$

where

$$w_{j-1} = \sum_{k \in Z} b_k^{j-1} \psi(2^{j-1}x - k) \in W_{j-1}$$

$$f_{j-1} = \sum_{k \in Z} a_k^{j-1} \phi(2^{j-1}x - k) \in V_{j-1}$$

with

$$b_k^{j-1} = \frac{a_{2k}^j - a_{2k+1}^j}{2} \qquad a_k^{j-1} = \frac{a_{2k}^j + a_{2k+1}^j}{2}.$$

The preceding process can now be repeated with j replaced by $j - 1$ to decompose f_{j-1} as $w_{j-2} + f_{j-2}$. Continuing in this way, we achieve the decomposition

$$f_j = w_{j-1} + w_{j-2} + \cdots + w_0 + f_0.$$

To summarize the decomposition process, a signal is first discretized that produces an approximate signal $f_j \in V_j$ as in Theorem 4.9. Then the decomposition algorithm in Theorem 4.12 produces a decomposition of f_j into its various frequency components: $f_j = w_{j-1} + w_{j-2} + \cdots + w_0 + f_0$.

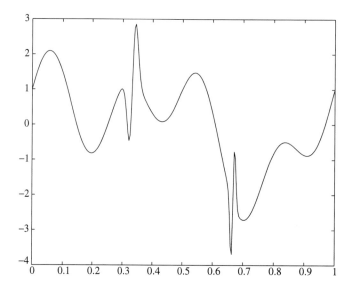

Figure 13 Sample signal

EXAMPLE 4.13

Consider the signal f defined on the unit interval $0 \le x \le 1$ given in Figure 13. We discretize this signal over 2^8 nodes [so $a_k^8 = f(k/2^8)$, $0 \le k \le 2^8 - 1$] since a mesh size of width $1/2^8$ seems to capture the essential features of this signal. Thus

$$f_8(x) = \sum_{k=0}^{2^8-1} f(k/2^8)\phi(2^8 x - k)$$

approximates f well enough for our purposes. Using Theorem 4.12, we decompose f_8 into its components in V_j, for $j = 8, 7, \ldots, 0$. Plots of the components in $f_8 \in V_8$, $f_7 \in V_7$, $f_6 \in V_6$, and $f_4 \in V_4$ are given in Figures 14, 16, 17, and 18 on the following three pages. A plot of the W_7-coefficients is given in Figure 15. The W_7-coefficients are small except where the original signal contains a sharp spike of width $\approx 2^{-8}$ (at $x \approx 0.3$ and $x \approx 0.65$). $\blacksquare$

4.3.2   Reconstruction

Having decomposed f into its V_0- and $W_{j'}$-components for $0 \le j' < j$, then what? The answer depends on the goal. If the goal is to filter out noise, then the $W_{j'}$-components of f corresponding to the unwanted frequencies can be thrown out and the resulting signal will have much less noise. If the goal is data compression, the $W_{j'}$-components that are small can be thrown out without appreciably changing the signal. Only the significant $W_{j'}$-components (the larger $b_k^{j'}$) need to be transmitted, and significant data compression can

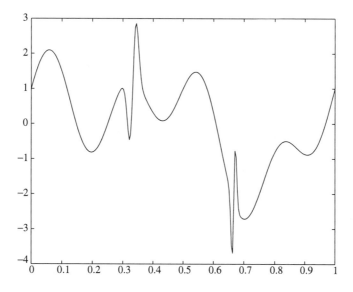

Figure 14 V_8-component

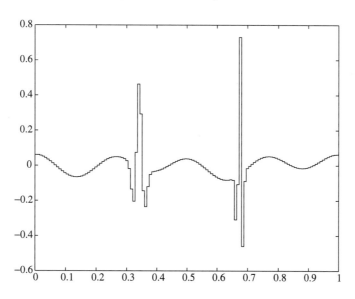

Figure 15 W_7-coefficients

be achieved. Of course, what constitutes "small" depends on the tolerance for error for a particular application.

In either case, since the $b_k^{j'}$ have been modified, we need a reconstruction algorithm (for the receiving end of the signal perhaps) so that the compressed or filtered signal can be rebuilt in terms of the basis elements $\phi(2^j x - l)$ of V_j;

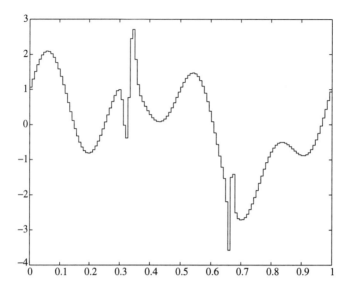

Figure 16 V_7-component

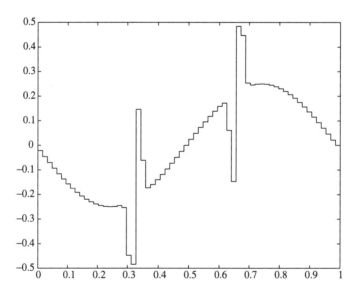

Figure 17 V_6-component

that is,

$$f(x) = \sum_{l \in Z} a_l^j \phi(2^j x - l).$$

Once this is done, the graph of the signal f is a step function of height a_l^j over the interval $l/2^j \leq x \leq (l+1)/2^j$.

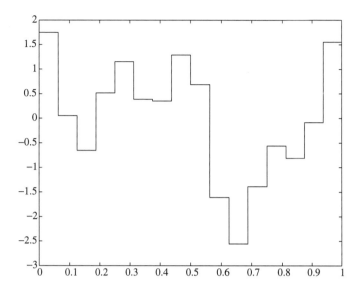

Figure 18 V_4-component

We start with a signal of the form

$$f(x) = f_0(x) + w_0(x) + \cdots + w_{j-1}(x) \qquad w_l \in W_l$$

where

$$f_0(x) = \sum_{k \in Z} a_k^0 \phi(x - k) \in V_0 \quad \text{and} \quad w_l = \sum_k b_k^l \psi(2^l x - k) \in W_l$$

for $0 \le l \le j - 1$. Our goal is to rewrite f as $f(x) = \sum_l a_l^j \phi(2^j x - l)$ and find an algorithm for the computation of the constants a_l^j. We use the equations

$$\phi(x) = \phi(2x) + \phi(2x - 1) \tag{4.12}$$
$$\psi(x) = \phi(2x) - \phi(2x - 1), \tag{4.13}$$

which follow from the definitions of ϕ and ψ. Replacing x by $2^{j-1}x$ yields

$$\phi(2^{j-1}x) = \phi(2^j x) + \phi(2^j x - 1) \tag{4.14}$$
$$\psi(2^{j-1}x) = \phi(2^j x) - \phi(2^j x - 1). \tag{4.15}$$

Using equation (4.12) with x replaced by $x - k$, we have

$$f_0(x) = \sum_{k \in Z} a_k^0 \phi(x - k) \quad \text{(the definition of } f_0\text{)}$$

$$= \sum_{k \in Z} a_k^0 \phi(2x - 2k) + a_k^0 \phi(2x - 2k - 1) \quad \text{from (4.12)}.$$

So

$$f_0(x) = \sum_{k \in Z} \hat{a}_l^1 \phi(2x - l) \qquad (4.16)$$

where

$$\hat{a}_l^1 = \begin{cases} a_k^0 & \text{if } l = 2k \text{ is even} \\ a_k^0 & \text{if } l = 2k + 1 \text{ is odd.} \end{cases}$$

Similarly, $w_0 = \sum_k b_k^0 \psi(x-k)$ can be written [using equation (4.13) for $\psi(x-k)$] as

$$w_0(x) = \sum_{l \in Z} \hat{b}_l^1 \phi(2x - l) \qquad (4.17)$$

where

$$\hat{b}_l^1 = \begin{cases} b_k^0, & \text{if } l = 2k \text{ is even} \\ -b_k^0, & \text{if } l = 2k + 1 \text{ is odd.} \end{cases}$$

Combining (4.16) and (4.17) yields

$$f_0(x) + w_0(x) = \sum_{l \in Z} a_l^1 \phi(2x - l)$$

where

$$a_l^1 = \hat{a}_l^1 + \hat{b}_l^1 = \begin{cases} a_k^0 + b_k^0, & \text{if } l = 2k \\ a_k^0 - b_k^0, & \text{if } l = 2k + 1. \end{cases}$$

Next, $w_1 = \sum_k b_k^1 \psi(2x - k)$ is added to this sum in the same manner [using (4.12) and (4.13) with x replaced by $2x - k$]:

$$f_0(x) + w_0(x) + w_1(x) = \sum_{l \in Z} a_l^2 \phi(2^2 x - l)$$

where

$$a_l^2 = \begin{cases} a_k^1 + b_k^1, & \text{if } l = 2k \\ a_k^1 - b_k^1, & \text{if } l = 2k + 1. \end{cases}$$

Note that the a_l^0- and b_l^0-coefficients determine the a_l^1-coefficients. Then the a_l^1- and b_l^1-coefficients determine the a_l^2-coefficients, and so on in a recursive manner.

The preceding reconstruction algorithm is summarized in the following theorem.

THEOREM 4.14 (Haar Reconstruction) *Suppose*

$$f = f_0 + w_0 + w_1 + w_2 + \cdots + w_{j-1}$$

with

$$f_0(x) = \sum_{k \in Z} a_k^0 \, \phi(x - k) \in V_0 \quad and \quad w_{j'}(x) = \sum_{k \in Z} b_k^{j'} \psi(2^{j'} x - k) \in W_{j'}$$

for $0 \le j' < j$. Then

$$f(x) = \sum_{l \in Z} a_l^j \, \phi(2^j x - l) \in V_j$$

where the $a_l^{j'}$ are determined recursively for $j' = 1$, then $j' = 2$, and so on until $j' = j$, by the algorithm

$$a_l^{j'} = \begin{cases} a_k^{j'-1} + b_k^{j'-1}, & if \ l = 2k \ is \ even \\ a_k^{j'-1} - b_k^{j'-1}, & if \ l = 2k + 1 \ is \ odd. \end{cases}$$

EXAMPLE 4.15

We apply the decomposition and reconstruction algorithms to compress the signal f that is shown in Figure 19; f is defined on the unit interval. (This is the same signal used in Example 4.13.)

We discretize this signal over 2^8 nodes [so $a_k^8 = f(k/2^8)$] and then decompose this signal (as in Theorem 4.12) to obtain $f = f_0 + w_0 + w_1 + w_2 + \cdots + w_7$ with

$$f_0(x) = a_0^0 \phi(x) \in V_0 \quad and \quad w_{j'}(x) = \sum_{k \in Z} b_k^{j'} \psi(2^{j'} x - k) \in W_{j'}$$

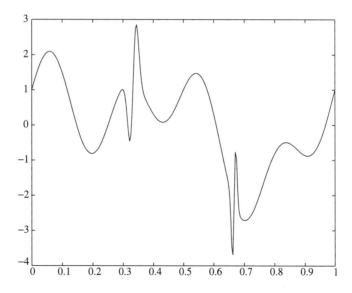

Figure 19 Sample signal

for $0 \leq j' \leq 7$. Note that there is only one basis term $\phi(x)$ in V_0 since the interval of consideration is $0 \leq x \leq 1$. We first use 85% compression on this decomposed signal, which means that after ordering the $|b_k^j|$ by size, we set the smallest 80% equal to zero (retaining the largest 20%). Then we reconstruct the signal as in Theorem 4.14. The result is graphed in Figure 20. Figure 21

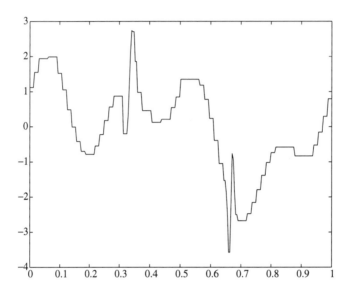

Figure 20 Eighty percent compression

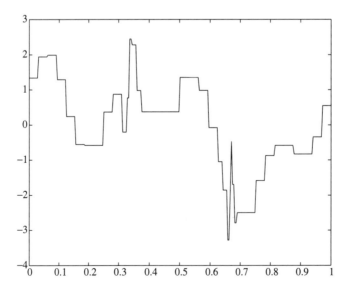

Figure 21 Ninety percent compression

illustrates the same process with 90% compression. The relative L^2-error is 0.0895 for 80% compression and 0.1838 for 90% compression. ■

4.3.3 Filters and Diagrams

The decomposition and reconstruction algorithms can be put in the language of discrete filters and simple operators acting on a sequence of coefficients. The algorithms can then be visualized in terms of block diagrams.

We do the decomposition algorithm first. Define two discrete filters (convolution operators) H and L via their impulse responses, which are the sequences h and ℓ:

$$h = (\cdots 0 \cdots \underbrace{-\frac{1}{2} \quad \frac{1}{2}}_{k=-1,\,0} \cdots 0 \cdots), \qquad \ell = (\cdots 0 \cdots \underbrace{\frac{1}{2} \quad \frac{1}{2}}_{k=-1,\,0} \cdots 0 \cdots).$$

If $\{x_k\} \in \ell^2$, then $H(x) := h * x$ and $L(x) := \ell * x$. The resulting sequences are thus

$$H(x)_k = (h * x)_k = \frac{1}{2}x_k - \frac{1}{2}x_{k+1} \qquad L(x)_k = (\ell * x)_k = \frac{1}{2}x_k + \frac{1}{2}x_{k+1}.$$

If we keep only even subscripts, then $H(x)_{2k} = (h * x)_{2k} = \frac{1}{2}x_{2k} - \frac{1}{2}x_{2k+1}$ and $L(x)_{2k} = (\ell * x)_{2k} = \frac{1}{2}x_{2k} + \frac{1}{2}x_{2k+1}$. This operation of discarding the odd coefficients in a sequence is called *downsampling*; we denote the corresponding operator by D.

We now apply these ideas to go from level-j scaling coefficients a_k^j to get the level $j - 1$ scaling and wavelet coefficients. Using Theorem 4.12 and replacing x by a_k^j,

$$b_k^{j-1} = DH(a^j)_k \quad \text{and} \quad a_k^{j-1} = DL(a^j)_k.$$

Figure 22 illustrates the decomposition algorithm. The downsampling operator D is replaced by the more suggestive symbol, "$2\downarrow$."

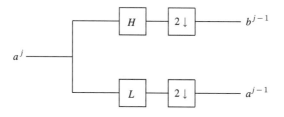

Figure 22 Haar decomposition diagram

The reconstruction algorithm also requires defining two discrete filters $\widetilde{H}$ and $\widetilde{L}$ via their corresponding impulse responses,

$$\widetilde{h} = (\cdots 0 \cdots \underbrace{1 \quad -1}_{k=0,\,1} \cdots 0 \cdots) \qquad \widetilde{\ell} = (\cdots 0 \cdots \underbrace{1 \quad 1}_{k=0,\,1} \cdots 0 \cdots).$$

For a sequence $\{x_k\}$, we have $(\widetilde{h} * x)_k = x_k - x_{k-1}$ and $(\widetilde{\ell} * x)_k = x_k + x_{k-1}$. Here is an important observation. If x and y are sequences in which the *odd* entries are all 0, then

$$(\widetilde{h} * x)_l = \begin{cases} x_{2k} & l = 2k \text{ is even,} \\ -x_{2k} & l = 2k+1 \text{ is odd,} \end{cases} \qquad (\widetilde{\ell} * y)_l = \begin{cases} y_{2k} & l = 2k \text{ is even,} \\ y_{2k} & l = 2k+1 \text{ is odd.} \end{cases}$$

Adding the two sequences $\widetilde{h} * x$ and $\widetilde{\ell} * y$ then gives us

$$(\widetilde{h} * x)_l + (\widetilde{\ell} * y)_l = \begin{cases} x_{2k} + y_{2k} & l = 2k \text{ is even,} \\ y_{2k} - x_{2k} & l = 2k+1 \text{ is odd.} \end{cases}$$

This is almost the pattern for the reconstruction formula given in Theorem 4.14. Although we have assumed that the x_{2k+1}'s and y_{2k+1}'s are 0, the x_{2k}'s and y_{2k}'s are ours to choose, so we set $x_{2k} = b_k^{j-1}$ and $y_{2k} = a_k^{j-1}$; that is,

$$x = (\cdots 0 \quad b_{-1}^{j-1} \quad 0 \quad \underbrace{b_0^{j-1}}_{k=0} \quad 0 \quad b_1^{j-1} \quad 0 \quad b_2^{j-1} \quad 0 \cdots)$$

and similarly for y. The sequences x and y are called *upsamples* of the sequences b^{j-1} and a^{j-1}. We use the U to denote the upsampling operator, so $x = Ub^{j-1}$ and $y = Ua^{j-1}$. The reconstruction formula in Theorem 4.14 then takes the compact form

$$a^j = \widetilde{L}Ua^{j-1} + \widetilde{H}Ub^{j-1}.$$

We illustrate the reconstruction step in Figure 23. The upsampling operator is replaced by the symbol $2\uparrow$.

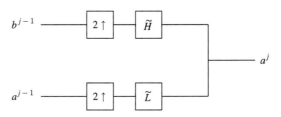

Figure 23 Haar reconstruction diagram

4.4 Summary

In this section, we present a summary of the ideas of this chapter. The format is a step-by-step procedure used to process (compress or denoise) given signal $y = f(t)$. We will let ϕ and ψ be the Haar scaling function and wavelet.

Step 1. *Sample.* If the signal is continuous (analog), $y = f(t)$, where t represents time, choose the top level $j = J$ so that 2^J is larger than the Nyquist rate for the signal (see the discussion just before the Theorem 2.23). Let

$$a_k^J = f(k/2^J).$$

In practice, the range of k is a finite interval determined by the duration of the signal. For example, if the duration of the signal is $0 \le t \le 1$, then the range for k is $0 \le k \le 2^J - 1$ (or perhaps $1 \le l \le 2^J$). If the signal is discrete to start with (i.e., a sequence of numbers), then this step is unnecessary. The top level a_k^J is set equal to the kth term in the sampled signal, and 2^J is taken to be the sampling rate. In any case, we have the highest-level approximation to f given by

$$f_J(x) = \sum_{k \in Z} a_k^J \phi(2^J x - k)$$

Step 2. *Decomposition.* The decomposition algorithm decomposes f_J into

$$f_J = w_{J-1} + \cdots + w_{j-1} + f_{j-1} + \cdots + w_0 + f_0,$$

where

$$w_{j-1} = \sum_{l \in Z} b_l^{j-1} \psi(2^{j-1} x - l)$$

$$f_{j-1} = \sum_{l \in Z} a_l^{j-1} \phi(2^{j-1} x - l).$$

The coefficients b_l^{j-1} and a_l^{j-1} are determined from the a^j recursively by the algorithm

$$a_l^{j-1} = DL(a^j)_k \tag{4.18}$$
$$b_l^{j-1} = DH(a^j)_k \tag{4.19}$$

where H and L are the high- and low-pass filters from Section 4.3.3. When $j = J$, a_k^{J-1} and b_k^{J-1} are determined by a_k^J, which are the sampled signal values from Step 1. Then j becomes $J - 1$ and a_k^{J-2} and b_k^{J-2} are determined from a_k^{J-1}. Then j becomes $J - 2$, and so on, until the level reached is either satisfactory for some purpose or there are too few coefficients to continue. Unless otherwise stated, the decomposition algorithm will continue until the $j = 0$ level is reached.

Step 3. *Processing.* After decomposition, the signal is in the form

$$f_J(x) = \sum_{j=0}^{J-1} w_j + f_0 \tag{4.20}$$

$$= \sum_{j=0}^{J-1} \left(\sum_{k \in Z} b_l^j \, \psi(2^j x - k) \right) + \sum_{k \in Z} a_k^0 \, \phi(x - k). \tag{4.21}$$

The signal can now be filtered by modifying the wavelet coefficients b_k^j. How this is to be done depends on the problem at hand. To filter out all high frequencies, all the b_k^j would be set to zero for j above a certain value. Perhaps only a certain segment of the signal corresponding to particular values of k is to be filtered. If data compression is the goal, then the b_k^j that are below a certain threshold (in absolute value) would be set to zero. Whatever the goal, the process modifies the b_k^j.

Step 4. *Reconstruction.* Now the goal is to take the modified signal, f_J, in the form (4.21) (with the modified b_k^j) and reconstruct it as

$$f_J = \sum_{k \in Z} a_k^J \phi(2^J x - k).$$

This is accomplished by the reconstruction algorithm discussed in Section 4.3.3:

$$a^j = \tilde{L}U a^{j-1} + \tilde{H}U b^{j-1} \tag{4.22}$$

for $j = 1, \ldots J$. When $j = 1$, the a_k^1 are computed from the a_k^0 and b_k^0. When $j = 2$, the a_k^2 are computed from the a_k^1 and b_k^1 and so forth. The range of k is determined by the time duration of the signal. When $j = J$ (the top level), a_k^J represents the approximate value of the processed signal at time $x = k/2^J$. Of course, these a_k^J are different from the original a_k^J due the modification of coefficients in Step 3.

4.5 Exercises

1. Let ϕ and ψ be the Haar scaling and wavelet functions, respectively. Let V_j and W_j be the spaces generated by $\phi(2^j x - k)$, $k \in Z$ and $\psi(2^j x - k)$, $k \in Z$, respectively. Consider the function defined on $0 \le x < 1$ given by

$$f(x) = \begin{cases} -1 & 0 \le x < 1/4 \\ 4 & 1/4 \le x < 1/2 \\ 2 & 1/2 \le x < 3/4 \\ -3 & 3/4 \le x < 1. \end{cases}$$

Express f first in terms of the basis for V_2 and then decompose f into its component parts in W_1, W_0, and V_0. In other words, find the Haar wavelet decomposition for f. Sketch each of these.

2. Repeat Exercise 1 for the function

$$f(x) = \begin{cases} 2 & 0 \le x < 1/4 \\ -3 & 1/4 \le x < 1/2 \\ 1 & 1/2 \le x < 3/4 \\ 3 & 3/4 \le x < 1. \end{cases}$$

3. If A and B are finite-dimensional, orthogonal subspaces of an inner product space V, then show

$$\dim(A \oplus B) = \dim A + \dim B.$$

If A and B are not necessarily orthogonal, then what is the relationship between $\dim(A + B)$, $\dim A$ and $\dim B$?

4. (a) Let V_n be the spaces generated by $\phi(2^n x - k)$, $k \in Z$, where ϕ is the Haar scaling function. On the interval $0 \leq x < 1$, what are the dimensions of the spaces V_n and W_n for $n \geq 0$ (just count the number of basis elements)?

   (b) Using the result of Exercise 3, count the dimension of the space on the right side of the equality

$$V_n = W_{n-1} \oplus W_{n-2} \oplus \cdots \oplus W_0 \oplus V_0.$$

   Is your answer the same as the one you computed for $\dim V_n$ in part (a)?

5. Let ϕ and ψ be the Haar scaling and wavelet functions, respectively. Let V_j and W_j be the spaces generated by $\phi(2^j x - k)$, $k \in Z$ and $\psi(2^j x - k)$, $k \in Z$, respectively. Suppose $f(x) = \sum_k a_k \phi(2x - k)$ ($a_k \in R$) belongs to V_1. Show explicitly that if f is orthogonal to each basis element $\phi(x - l) \in V_0$, for all integers l, then $a_{2l+1} = -a_{2l}$ for all integers l and hence show that

$$f(x) = \sum_{l \in Z} a_{2l} \psi(x - l) \in W_0.$$

6. Reconstruct $g \in V_3$ given these coefficients in its Haar wavelet decomposition:

$$a^2 = [1/2, 2, 5/2, -3/2] \quad b^2 = [-3/2, -1, 1/2, -1/2].$$

The first entry in each list corresponds to $k = 0$. Sketch g.

7. Reconstruct $h \in V_3$ given these coefficients in its Haar wavelet decomposition:

$$a^1 = [3/2, -1] \quad b^1 = [-1, -3/2] \quad b^2 = [-3/2, -3/2, -1/2, -1/2].$$

The first entry in each list corresponds to $k = 0$. Sketch h.

The remaining problems require some programming in a language such as MATLAB, *Maple, or C. The code in Appendix B entitled 'Matlab Code' should be useful.*

8. (Haar wavelets on $[0, 1]$.) Let n be an integer, and let f be a function continuous on $[0, 1]$. Let $h_k(t) = \sqrt{n}\phi(nt - k)$, where $\phi(t)$ is the Haar scaling

function, which is 1 for t in $[0, 1)$ and 0 otherwise, form the projection of f onto the span of the h_k's,

$$f_n = \langle f, h_0 \rangle h_0 + \cdots + \langle f, h_{n-1} \rangle h_{n-1}.$$

Show that f_n converges uniformly to f on $[0, 1]$. For $f(t) = 1 - t^2$, use MATLAB or Maple to find the Haar wavelet decomposition (on $[0, 1]$) for $n = 4$, 8, and 16. Plot the results.

9. Let
$$f(t) = e^{-t^2/10} \left(\sin(2t) + 2 \cos(4t) + 0.4 \sin(t)\sin(50t) \right).$$

Discretize the function f over the interval $0 \le t \le 1$ as described in Step 1 of Section 4.4. Use $n = 8$ as the top level (so there are 2^8 nodes in the discretization). Implement the decomposition algorithm described in Step 2 of Section 4.4 using the Haar wavelets. Plot the resulting levels, $f_{j-1} \in V_{j-1}$ for $j = 8 \ldots 1$, and compare with the original signal.

10. (Continuation of Exercise 9). Filter the wavelet coefficients computed in Exercise 9 by setting to zero any wavelet coefficient whose absolute value is less than $tol = 0.1$. Then reconstruct the signal as described in Section 4.4. Plot the reconstructed f_8 and compare with the original signal. Compute the relative l^2-difference between the original signal and the compressed signal. Experiment with various tolerances. Keep track of the percentage of wavelet coefficients that have been filtered out.

11. Haar wavelets can be used to detect a discontinuity in a signal. Let $g(t)$ be defined on $0 \le t < 1$ via

$$g(t) = \begin{cases} 0 & 0 \le t < 7/17 \\ 1 - t^2 & 7/17 \le t < 1. \end{cases}$$

Discretize the functions g over the interval $0 \le t \le 1$ as described in Step 1 of Section 4.4. Use $n = 7$ as the top level (so there are 2^7 nodes in the discretization). Implement a one-level decomposition. Plot the magnitudes of the level 6 wavelet *coefficients*. Which wavelet has the largest coefficient? What t corresponds to this wavelet? Try the method again with $7/17$ relaced by $8/9$, and then by $2/7$. Why do you think the method works?

Chapter 5

Multiresolution Analysis

In the previous chapter, we described a procedure for decomposing a signal into its Haar wavelet components of varying frequencies (see Theorem 4.12). The Haar wavelet scheme relied on two functions: the Haar scaling function ϕ and the Haar wavelet ψ. Both are simple to describe and lead to an easy decomposition algorithm. The drawback with the Haar decomposition algorithm is that both of these functions are discontinuous (ϕ at $x = 0$, 1 and ψ at $x = 0, 1/2, 1$). As a result, the Haar decomposition algorithm provides only crude approximations to a continuously varying signal (as already mentioned, Figure 3 in Chapter 4 does not approximate Figure 2 in Chapter 4 very well). What is needed is a theory similar to what has been described in the past sections but with continuous versions of our building blocks, ϕ and ψ. In this chapter, we present a framework for creating more general ϕ and ψ. The resulting theory, due to Stéphane Mallat [10, 11], is called a *multiresolution analysis*. In the sections that follow, this theory will be used together with a continuous ϕ and ψ (to be constructed later) that will improve the performance of the signal decomposition algorithm with the Haar wavelets described in the past section.

5.1 The Multiresolution Framework

5.1.1 Definition

Before we define the notion of a multiresolution analysis, we need to provide some background. Recall that the sampling theorem (Theorem 2.23) approximately reconstructs a signal f from samples taken uniformly at intervals of length T. If the signal is band limited and its Nyquist frequency is less than $1/T$, then the reconstruction is perfect; otherwise it's an approximation. The smaller T is, the better we can approximate or *resolve* the signal; the size of T measures our *resolution* of the signal f. A typical FFT analysis of the samples taken from f works at *one* resolution, T.

If the signal has bursts where it varies rapidly, interspersed with periods where it is slowly varying, this "single" resolution analysis does not work well— for all the reasons that we outlined earlier in Section 4.1. To treat such signals, Mallat had the idea to do these two things. First, replace the space of band-limited functions from the sampling theorem with one tailored to the signal. Second, analyze the signal using the scaled versions of the same space, but geared to resolutions $T/2$, $T/2^2$, and so on (hence the term *multi*resolution analysis). This framework is ideal for not only analyzing certain signals, but also for actually creating wavelets.

We start with the general definition of a multiresolution analysis.

DEFINITION 5.1   *Let V_j , $j = \ldots, -2, -1, 0, 1, 2, \ldots$ be a sequence of sub-spaces of functions in $L^2(R)$. The collection of spaces $\{V_j, j \in Z\}$ is called a multiresolution analysis with scaling function ϕ if the following conditions hold:*

1. *(nested) $V_j \subset V_{j+1}$*

2. *(density) $\overline{\cup V_j} = L^2(R)$*

3. *(separation) $\cap V_j = \{0\}$*

4. *(scaling) The function $f(x)$ belongs to V_j if and only if the function $f(2^{-j}x)$ belongs to V_0.*

5. *(orthonormal basis) The function ϕ belongs to V_0 and the set $\{\phi(x-k), k \in Z\}$ is an orthonormal basis (using the L^2 inner product) for V_0.*

The V_j's are call *approximation spaces*. There may be several choices of ϕ corresponding to a system of approximation spaces. Different choices for ϕ may yield different multiresolution analyses. Although we required the translates of $\phi(x)$ to be orthonormal, we don't have to. All that is needed is a ϕ for which the set $\{\phi(x - k), k \in Z\}$ is a basis. We can then use ϕ to obtain a new scaling function $\tilde{\phi}$ for which $\{\tilde{\phi}(x - k), k \in Z\}$ *is* orthonormal. We discuss this in Section 5.3.

Probably the most useful class of scaling functions are those that have compact or finite support. As defined in the previous chapter, a function has compact support if it is identically 0 outside of a finite interval. The Haar scaling function is a good example of a compactly supported function. The scaling functions associated with Daubechies's wavelets are not only compactly supported, but also continuous. Having both properties in a scaling function is especially desirable, because the associated decomposition and reconstruction algorithms are computationally faster and do a better job of analyzing and reconstructing signals.

EXAMPLE 5.2

The Haar scaling function (Definition 4.1) generates the subspaces, V_j, consisting of the space of piecewise constant functions of finite support (i.e., step functions) whose discontinuities are contained in the set of integer multiples of 2^{-j} (see Definition 4.4). We now verify that this collection $\{V_j, \; j \geq 0\}$ together with the Haar scaling function, ϕ, satisfies the definition of a multiresolution analysis.

As mentioned just after Definition 4.4, the nested property follows since the set of multiples of 2^{-j} is contained in the set of multiples of $2^{-(j+1)}$ (the former set consists of every other member of the latter set). Intuitively, an approximation of a signal by a function in V_j is capable of capturing details of the signal down to a resolution of 2^{-j}. The nested property indicates that as j gets larger, more information is revealed by an approximation of a signal by a function in V_j.

The density condition for the Haar system is the content of Theorem 4.9. This condition means that an approximation of a signal by a function in V_j eventually captures all details of the signal as j gets larger (i.e., as $j \to \infty$). This approximation is illustrated in Figure 11.

To discuss the separation condition, first note that j can be negative as well as positive in the definition of V_j. If f belongs to V_{-j}, for $j > 0$, then f must be a finite linear combination of $\{\phi(x/2^j - k), k \in Z\}$ whose elements are constant on the intervals $\ldots, [-2^j, 0), [0, 2^j), \ldots$. As j gets larger, these intervals get larger. On the other hand, the support of f (i.e., the set where f is nonzero) must stay finite. So if f belongs to all the V_{-j} as $j \to \infty$, then these constant values of f must be zero.

Finally, the scaling condition for the Haar system follows from Theorem 4.5 and the orthonormal condition follows from Theorem 4.6. Therefore, the Haar system of $\{V_j\}$ satisfies all the properties of a multiresolution analysis. ∎

EXAMPLE 5.3

Linear Spline Multiresolution Analysis Linear splines are continuous, piecewise linear functions; the jagged function in Figure 1 is an example of one.

We can construct a multiresolution analysis for linear splines. Let V_j be the space of all finite energy signals f that are continuous and piecewise linear, with

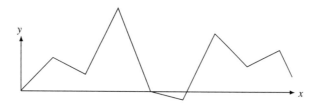

Figure 1 A linear spline

possible corners occurring only at the dyadic points $k/2^j$, $k \in Z$. These approximation spaces satisfy the conditions 1–4 in the definition of a multiresolution analysis (see Exercise 9). The scaling function φ is the "tent function,"

$$\varphi(x) = \begin{cases} x+1, & -1 \le x \le 0, \\ 1-x, & 0 < x \le 1, \\ 0, & |x| > 1. \end{cases}$$

and the set $\{\varphi(x-k)\}_{k \in Z}$ is a *non*orthogonal one. Using the methods in Section 5.3, we can construct a new scaling function ϕ that does generate an orthonormal basis for V_0.   ■

EXAMPLE 5.4
_____

Shannon Multiresolution Analysis   For $j \in Z$, let V_j be the space of all finite energy signals f for which the Fourier transform $\hat{f} = 0$ outside of the interval $[-2^j\pi, 2^j\pi]$—that is, all $f \in L^2(R)$ that are band limited and have $\text{supp}(\hat{f}) \subseteq [-2^j\pi, 2^j\pi]$. The scaling function is $\phi(x) = \text{sinc}(x)$, where

$$\text{sinc}(x) := \begin{cases} 1 & x = 0 \\ \dfrac{\sin(\pi x)}{\pi x} & x \ne 0. \end{cases} \tag{5.1}$$

See Exercise 8 for further details.   ■

We turn to a discussion of properties common to every multiresolution analysis. For a given function $g : R \to R$, let

$$g_{jk}(x) = 2^{j/2} g(2^j x - k).$$

The function $g_{jk}(x) = 2^{j/2} g(2^j(x - k/2^j))$ is a translation (by $k/2^j$) and a rescaling (by a factor of $2^{j/2}$) of the original function g. (See Exercise 1.) The factor of $2^{j/2}$ is present to preserve its L^2 norm; that is,

$$\|g_{jk}\|_{L^2}^2 = \|g\|_{L^2}^2 \quad \text{for all } j, k.$$

We use this notation both for the scaling function ϕ and, later, for the wavelet. Our first result is that $\{\phi_{jk}\}_{k \in Z}$ is an orthonormal basis for V_j.

THEOREM 5.5   *Suppose $\{V_j; j \in Z\}$ is a multiresolution analysis with scaling function ϕ. Then for any $j \in Z$, the set of functions*

$$\{\phi_{jk}(x) = 2^{j/2}\phi(2^j x - k); \ k \in Z\}$$

is an orthonormal basis for V_j.

Proof To prove that $\{\phi_{jk}, k \in Z\}$ spans V_j, we must show that any $f(x) \in V_j$ can be written as a linear combination $\{\phi(2^j x - k); \ k \in Z\}$. Using the scaling property (Definition 5.1, condition 4), the function $f(2^{-j}x)$ belongs to V_0 and therefore $f(2^{-j}x)$ is a linear combination of $\{\phi(x - k), k \in Z\}$. By replacing x by $2^j x$, we see that $f(x)$ is a linear combination of $\{\phi(2^j x - k), k \in Z\}$.

To show that $\{\phi_{jk}, k \in Z\}$ is orthonormal, we must show

$$\langle \phi_{jk}, \phi_{jl} \rangle_{L^2} = \delta_{jk} = \begin{cases} 0 & \text{if } j \neq k \\ 1 & \text{if } j = k \end{cases}$$

or $\quad 2^j \displaystyle\int_{-\infty}^{\infty} \phi(2^j x - k)\overline{\phi(2^j x - l)}\,dx = \delta_{kl}.$

To establish this equation, make the change of variables $y = 2^j x$ (and so $dy = 2^j dx$). We obtain

$$2^j \int_{-\infty}^{\infty} \phi(2^j x - k)\overline{\phi(2^j x - l)}\,dx = \int_{-\infty}^{\infty} \phi(y - k)\overline{\phi(y - l)}\,dy,$$

which equals δ_{kl} in view of the orthonormal basis property given in Definition 5.1, condition 5. $\qquad\blacklozenge$

5.1.2 The Scaling Relation

We are now ready to state and prove the central equation in multiresolution analysis, the *scaling relation*.

THEOREM 5.6 *Suppose $\{V_j; j \in Z\}$ is a multiresolution analysis with scaling function ϕ. Then the following scaling relation holds:*

$$\phi(x) = \sum_{k \in Z} p_k \phi(2x - k) \quad \text{where} \quad p_k = 2 \int_{-\infty}^{\infty} \phi(x)\overline{\phi(2x - k)}\,dx.$$

Moreover, we also have

$$\phi(2^{j-1}x - l) = \sum_{k \in Z} p_{k-2l}\phi(2^j x - k) \tag{5.2}$$

or, equivalently,

$$\phi_{j-1,l} = 2^{-1/2} \sum_{k} p_{k-2l}\phi_{jk} \tag{5.3}$$

where $\phi_{jk}(x) = 2^{j/2}\phi(2^j x - k)$.

Remark. The preceding equation, which relates $\phi(x)$ and the translates of $\phi(2x)$, is called the *scaling relation* or *two-scale relation*. When the support of ϕ is compact, only a finite number of the p_k are nonzero, because when $|k|$ is large enough, the support of $\phi(2x - k)$ will be outside of the support for $\phi(x)$. Therefore, the sum occurring in Theorem 5.6 is finite. Usually, ϕ is real valued in which case the p_k are real.

Proof   To prove this theorem, note that $\phi(x) = \sum \tilde{p}_k \phi_{1k}(x)$ must hold for some choice of $\tilde{p}_k$ because $\phi(x)$ belongs to $V_0 \subset V_1$, which is the linear span of $\{\phi_{1k}, k \in Z\}$. Since $\{\phi_{1k}, k \in Z\}$ is an orthonormal basis of V_1, the $\tilde{p}_k$ can be determined by using Theorem 0.21:

$$\tilde{p}_k = \langle \phi, \phi_{1k} \rangle_{L^2} = 2^{1/2} \int_{-\infty}^{\infty} \phi(x) \overline{\phi(2x - k)} \, dx.$$

Therefore ,

$$\phi(x) = \sum_{k \in Z} \tilde{p}_k \phi_{1k}(x) = \sum_{k \in Z} \tilde{p}_k 2^{1/2} \phi(2x - k).$$

Let $p_k = 2^{1/2} \tilde{p}_k = 2 \int_{-\infty}^{\infty} \phi(x) \, \overline{\phi(2x - k)} \, dx$ and we have

$$\phi(x) = \sum_{k \in Z} p_k \phi(2x - k)$$

as claimed. To get the second equation, replace x by $2^{j-1}x - l$ in ϕ, and then adjust the index of summation in the resulting series. The third follows from the second by multiplying by $2^{(j-1)/2}$ and then simplifying.   ♦

EXAMPLE 5.7

The values of the p_k for the Haar system are

$$p_0 = p_1 = 1$$

[see (4.12)] and all other p_k are zero.   ■

EXAMPLE 5.8

As mentioned earlier, the Haar scaling function and Haar wavelet are discontinuous. There is a more intricate construction, due to Daubechies, that we will present in Chapter 6 that leads to a continuous scaling function and wavelet (see Figures 2 and 3 in Chapter 6 for their graphs). The associated values of the p_k will be computed in Section 6.1 to be

$$p_0 = \frac{1 + \sqrt{3}}{4} \quad p_1 = \frac{3 + \sqrt{3}}{4} \quad p_2 = \frac{3 - \sqrt{3}}{4} \quad p_3 = \frac{1 - \sqrt{3}}{4}. \tag{5.4}$$

■

The following result contains identities for the p_k that will be important later.

THEOREM 5.9   *Suppose $\{V_j; j \in Z\}$ is a multiresolution analysis with scaling function ϕ. Then the following equalities hold:*

1. $\displaystyle\sum_{k \in Z} p_{k-2l} \overline{p_k} = 2\delta_{l0}$

2. $\displaystyle\sum_{k \in Z} |p_k|^2 = 2$

3. $\displaystyle\sum_{k \in Z} p_k = 2$

4. $\displaystyle\sum_{k \in Z} p_{2k} = 1 \quad and \quad \sum_{k \in Z} p_{2k+1} = 1$

Proof The first equation follows from the two-scale relation (Theorem 5.6) and the fact that $\{\phi(x - k), k \in Z\}$ is orthonormal. We leave the details to Exercise 4a. By setting $l = 0$ in the first equation, we get the second equation.

The proof of the third equation uses Theorem 5.6 as follows:

$$\int_{-\infty}^{\infty} \phi(x)\,dx = \sum_{k \in Z} p_k \int_{-\infty}^{\infty} \phi(2x - k)\,dx.$$

By letting $t = 2x - k$ and $dx = dt/2$, we obtain

$$\int_{-\infty}^{\infty} \phi(x)\,dx = 1/2 \sum_{k \in Z} p_k \int_{-\infty}^{\infty} \phi(t)\,dt.$$

Now $\int \phi(t)\,dt$ cannot equal zero (see Exercise 6), for otherwise we could never approximate functions $f \in L^2(R)$ with $\int f(t)\,dt \neq 0$ by functions in V_j. Therefore, the factor of $\int \phi(t)\,dt$ on the right can be canceled with $\int \phi(x)\,dx$ on the left. The third equation now follows.

To prove the fourth equation, we take the first equation, $\sum_k p_{k-2l}\bar{p}_k = 2\delta_{l0}$, and replace l by $-l$ and then sum over l. We obtain

$$\sum_{l \in Z} \sum_{k \in Z} p_{k+2l}\bar{p}_k = 2 \sum_{l \in Z} \delta_{l0} = 2.$$

Now divide the sum over k into even and odd terms:

$$2 = \sum_{l \in Z} \left(\sum_{k \in Z} p_{2k+2l}\,\bar{p}_{2k} + \sum_{k \in Z} p_{2k+1+2l}\,\bar{p}_{2k+1} \right)$$

$$= \sum_{k \in Z} \left(\sum_{l \in Z} p_{2k+2l} \right) \bar{p}_{2k} + \sum_{k \in Z} \left(\sum_{l \in Z} p_{2k+1+2l} \right) \bar{p}_{2k+1}.$$

For the two inner l-sums on the right, replace l by $l - k$. These inner l-sums become $\sum_l p_{2l}$ and $\sum_l p_{2l+1}$, respectively. Therefore,

$$2 = \sum_{k \in Z} \bar{p}_{2k} \sum_{l \in Z} p_{2l} + \sum_{k \in Z} \bar{p}_{2k+1} \sum_{l \in Z} p_{2l+1}$$

$$= |\sum_{k \in Z} \bar{p}_{2k}|^2 + |\sum_{k \in Z} \bar{p}_{2k+1}|^2.$$

Let $E = \sum_{k \in Z} p_{2k}$ and $O = \sum_{k \in Z} p_{2k+1}$. This last equation can be restated as $|E|^2 + |O|^2 = 2$. From the first property, $\sum_k p_k = 2$. Dividing this sum into even and odd indices gives $E + O = 2$. These two equations for E and O have only one solution: $E = O = 1$, as illustrated in Figure 2 for the case when all the p_k are real numbers. For the general case, see Exercise 7.   ◆

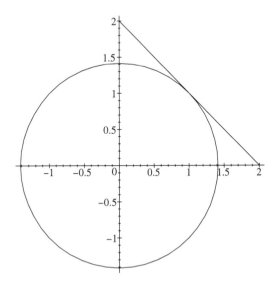

Figure 2 The circle $E^2 + O^2 = 2$ and the line $E + O = 2$

5.1.3   The Associated Wavelet and Wavelet Spaces

Recall that V_j is a subset of V_{j+1}. In order to carry out the decomposition algorithm in the general case, we need to decompose V_{j+1} into an orthogonal direct sum of V_j and its orthogonal complement, which we denote by W_j (as we did in the case of the Haar system). In addition, we need to construct a function ψ whose translates generate the space W_j (also as we did in the Haar system). Once ϕ is specified, the scaling relation can be used to construct the associated function ψ that generates W_j. We do this now.

THEOREM 5.10   *Suppose $\{V_j; j \in Z\}$ is a multiresolution analysis with scaling function*

$$\phi(x) = \sum_k p_k \phi(2x - k)$$

(p_k are the coefficients in Theorem 5.6). Let W_j be the span of $\{\psi(2^j x - k); k \in Z\}$, where

$$\psi(x) = \sum_{k \in Z} (-1)^k \overline{p_{1-k}} \phi(2x - k). \tag{5.5}$$

Then $W_j \subset V_{j+1}$ is the orthogonal complement of V_j in V_{j+1}. Furthermore, $\{\psi_{jk}(x) := 2^{j/2}\psi(2^j x - k), k \in Z\}$ is an orthonormal basis for the W_j.

For the Haar scaling function, the coefficients p_0 and p_1 are both 1. Theorem 5.10 then states that the Haar wavelet is $\psi(x) = \phi(2x) - \phi(2x-1)$, which agrees with the definition of ψ in Definition 4.7.

Proof of Theorem 5.10 Once we establish this theorem for the case $j = 0$, the case for $j > 0$ will follow by using the scaling condition (Definition 5.1, condition 4).

To establish the result for the case $j = 0$, we must verify the following three statements:

1. The set $\{\psi_{0k}(x) = \psi(x - k), \ k \in Z\}$ is orthonormal.

2. $\psi_{0k}(x) = \psi(x - m)$ is orthogonal to V_0 for all $m \in Z$. Equivalently, W_0 is contained in the orthogonal complement of V_0 in V_1.

3. Any function in V_1 that is orthogonal to V_0 can be written as a linear combination of $\psi(x - k)$. Equivalently, W_0 contains the orthogonal complement of V_0 in V_1.

We shall establish the first two statements and postpone the proof of third (which is more technical) until Appendix A. The second and third together imply that W_0 *is* the orthogonal complement of V_0 in V_1.

To establish the first statement we have from Exercise 4b that

$$\langle \psi_{0m}, \psi_{0l} \rangle = \frac{1}{2} \sum_{k \in Z} \overline{p_{1-k+2m}} \, p_{1-k+2l}.$$

Making the change of index $k' = 1 - k + 2m$ in this series and using the first part of Theorem 5.9, we have

$$\langle \psi_{0m}, \psi_{0l} \rangle = \frac{1}{2} \sum_{k' \in Z} \overline{p_{k'}} \, p_{k'+2l-2m} = \delta_{m-l,0} = \delta_{m,l},$$

so $\psi_{0,m}$ is orthonormal.

To establish the second statement, it is sufficient to show that $\psi(x - m)$ is orthogonal to $\phi(x - l)$ for each $l \in Z$ because V_0 is spanned by $\{\phi(x-l), \ l \in Z\}$. From Exercise 4c, we have that

$$\langle \phi_{0l}, \psi_{0m} \rangle = \frac{1}{2} \sum_{k \in Z} (-1)^k p_{1-k+2m} \, p_{k-2l}.$$

The sum on the right is zero. To see this, let us first examine the case when $l = m = 0$, in order to see the pattern of cancellation. In this case, this sum is

$$\sum_{k \in Z} (-1)^k p_{1-k} p_k = \cdots - p_2 p_{-1} + p_1 p_0 - p_0 p_1 + p_{-1} p_2 + \cdots.$$

The inner pair of terms cancel and then the outer pair of terms cancel, and so on. To see the general case, first note that the term with $k = l + m - j$ for $j \geq 0$ is

$$(-1)^{l+m-j} p_{1-l+m+j} p_{m-l-j},$$

which cancels with the term with $k = l + m + j + 1$ (again with $j \geq 0$), which is

$$(-1)^{l+m+j+1} p_{m-l-j} p_{m+j+1-l} = -(-1)^{l+m-j} p_{m-l-j} p_{m+j+1-l}.$$

This completes the proof of the second statement. As we noted earlier, the third statement will be proved in Appendix A.   ♦

For future reference, we note that $\psi_{jl}(x) = 2^{j/2}\psi(2^j x - l)$ has the expansion

$$\psi_{jl} = 2^{-1/2} \sum_{k \in Z} (-1)^k \overline{p_{1-k+2l}} \phi_{j+1,k}. \tag{5.6}$$

This follows from the definition of ψ given in Theorem 5.10: Substitute $2^j x - l$ for x, multiply both both sides by $2^{j/2}$, and adjust the summation index.

From Theorem 5.10, the set $\{\psi_{j-1,k}\}_{k \in Z}$ is an orthonormal basis for the space W_{j-1}, which is the orthogonal complement of V_{j-1} in V_j (so $V_j = W_{j-1} \oplus V_{j-1}$). By successive orthogonal decompositions,

$$\begin{aligned}
V_j &= W_{j-1} \oplus V_{j-1} \\
&= W_{j-1} \oplus W_{j-2} \oplus V_{j-2} \\
&\cdots \\
&= W_{j-1} \oplus W_{j-2} \oplus \cdots \oplus W_0 \oplus V_0.
\end{aligned}$$

Since we have defined V_j for $j < 0$, we can keep going:

$$\begin{aligned}
V_j &= W_{j-1} \oplus W_{j-2} \oplus \cdots \oplus W_0 \oplus W_{-1} \oplus V_{-1} \\
&\cdots \\
&= W_{j-1} \oplus W_{j-2} \oplus \cdots \oplus W_{-1} \oplus W_{-2} \cdots .
\end{aligned}$$

The V_j are nested and the union of all the V_j is the space $L^2(R)$. Therefore, we can let $j \to \infty$ and obtain the following theorem.

THEOREM 5.11 *Let $\{V_j, j \in Z\}$ be a multiresolution analysis with scaling function ϕ. Let W_j be the orthogonal complement of V_j in V_{j+1}. Then*

$$L^2(R) = \cdots \oplus W_{-1} \oplus W_0 \oplus W_1 \oplus \cdots .$$

In particular, each $f \in L^2(R)$ can be uniquely expressed as a sum $\sum_{k=-\infty}^{\infty} w_k$ with $w_k \in W_k$ and where the w_k's are mutually orthogonal. Equivalently, the set of all wavelets, $\{\psi_{jk}\}_{j,k \in Z}$, is an orthonormal basis for $L^2(R)$.

The infinite sum appearing in this theorem should be thought of as an approximation by finite sums. In other words, each $f \in L^2(R)$ can be approximated arbitrarily closely in the L^2-norm by finite sums of the form $w_{-j} + w_{1-j} + \cdots + w_{j-1} + w_j$ for suitably large j.

For large j, the W_j-components of a signal represent its high-frequency components because W_j is generated by translates of the function $\psi(2^j x)$ that vibrate at high frequency. For example, compare the picture in Figure 3, which is the graph of a function, $\psi(x)$, to that in Figure 4, which is the graph of $\psi(2^2 x)$ (the same function ψ is used to generate both graphs).

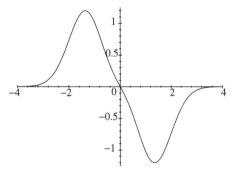

Figure 3 Graph of ψ

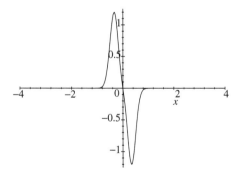

Figure 4 Graph of $\psi(2^2 x)$

5.1.4   Decomposition and Reconstruction Formulas: A Tale of Two Bases

Suppose that we are dealing with a signal f that is already in one of the approximation space, such as V_j. There are two primary orthonormal bases that can be used for representing f. The first is the natural scaling function basis for V_j, $\{\phi_{jk}\}_{k \in Z}$ as defined in Theorem 5.5. In terms of this basis, we have

$$f = \sum_{k \in Z} \langle f, \phi_{jk} \rangle \phi_{jk} \,. \tag{5.7}$$

Of course, since we have the orthogonal direct sum decomposition $V_j = V_{j-1} \oplus W_{j-1}$, we can also use the concatenation of the bases for V_{j-1} and W_{j-1}; that is, we use $\{\phi_{j-1,k}\}_{k \in Z} \cup \{\psi_{j-1,k}\}_{k \in Z}$, where ψ_{jk} is defined in Theorem 5.10. Relative to this orthonormal basis (see Theorem 0.21), f has the form

$$f = \underbrace{\sum_{k \in Z} \langle f, \phi_{j-1,k} \rangle \phi_{j-1,k}}_{f_{j-1}} + \underbrace{\sum_{k \in Z} \langle f, \psi_{j-1,k} \rangle \psi_{j-1,k}}_{w_{j-1}} \,. \tag{5.8}$$

The *decomposition* formula starts with the coefficients relative to the first basis [in (5.7)] and uses them to calculate the coefficients relative to the second basis [in (5.8)]. The *reconstruction* formula does the reverse. Here is the decomposition formula:

Decomposition:
$$\begin{cases} \langle f, \phi_{j-1,l} \rangle = 2^{-1/2} \sum_{k \in Z} \overline{p_{k-2l}} \langle f, \phi_{jk} \rangle \\ \langle f, \psi_{j-1,l} \rangle = 2^{-1/2} \sum_{k \in Z} (-1)^k p_{1-k+2l} \langle f, \phi_{jk} \rangle. \end{cases} \tag{5.9}$$

The two parts of this formula are obtained by using Parseval's equation (Exercise 3) in conjunction with appropriate expansions. To obtain the first part, we use the expansion (5.7) for f and the scaling relation in equation (5.3). The second part also requires (5.7), but uses equation (5.6) for the wavelets ψ_{jk}, with j replaced by $j - 1$.

One important application of the decomposition formula is to express ϕ_{jk} in terms of the second basis. From the decomposition formula (5.9) together with the orthonormality of the ϕ_{jk}'s, we have that $\langle \phi_{jk}, \phi_{j-1,l} \rangle = 2^{-1/2} \overline{p_{k-2l}}$ and $\langle \phi_{jk}, \psi_{j-1,l} \rangle = 2^{-1/2}(-1)^k p_{1-k+2l}$. Being careful about which index we are summing over (l in this case), we obtain the expansion

$$\phi_{jk} = \sum_{l \in Z} 2^{-1/2} \overline{p_{k-2l}} \phi_{j-1,l} + \sum_{l \in Z} 2^{-1/2}(-1)^k p_{1-k+2l} \psi_{j-1,l} \,, \tag{5.10}$$

which is literally the "inverse" of the scaling relation.

If we apply Parseval's equation (Exercise 3) to equations (5.10) and (5.8), we obtain the reconstruction formula:

Reconstruction:
$$\begin{cases} \langle f, \phi_{jk} \rangle = 2^{-1/2} \sum_{l \in Z} p_{k-2l} \langle f, \phi_{j-1,l} \rangle \\ \qquad\qquad + 2^{-1/2} \sum_{l \in Z} (-1)^k \overline{p_{1-k+2l}} \langle f, \psi_{j-1,l} \rangle. \end{cases} \tag{5.11}$$

The preceding formulas all involve *orthonormal* bases. For various reasons it might be more convenient to use the orthogonal versions—$\{\phi(2^j x - k)\}_{k \in Z}$—rather than $\{\phi_{jk}(x) = 2^{j/2}\phi(2^j x - k)\}_{k \in Z}$, and $\{\psi(2^j x - k)\}_{k \in Z}$ rather than

$\{\psi_{jk}(x) = 2^{j/2}\psi(2^j x - k)\}_{k \in Z}$. For instance, the expansion for $f \in V_j$ in equation (5.7) can be rewritten this way:

$$f(x) = \sum_{k \in Z} \underbrace{2^{j/2}\langle f, \phi_{jk}\rangle}_{a_k^j} \underbrace{2^{-j/2}\phi_{jk}}_{\phi(2^j x - k)}$$

$$= \sum_{k \in Z} a_k^j \phi(2^j x - k).$$

Similarly, equation (5.8) can be rewritten as

$$f = \sum_{k \in Z} a_k^{j-1}\phi(2^{j-1}x - k) + \sum_{k \in Z} b_k^{j-1}\psi(2^{j-1}x - k)$$

where $a_k^{j-1} = 2^{(j-1)/2}\langle f, \phi_{j-1,k}\rangle$ and $b_k^{j-1} = 2^{(j-1)/2}\langle f, \psi_{j-1,k}\rangle$. The a_k^j's are called the *approximation coefficients* and the b_k^j's are called the *detail coefficients*. The decomposition and reconstruction formulas can be restated in terms of these coefficients.

Decomposition: $\begin{cases} a_l^{j-1} = 2^{-1}\sum_{k \in Z} \overline{p_{k-2l}}a_k^j \\ b_l^{j-1} = 2^{-1}\sum_{k \in Z}(-1)^k p_{1-k+2l}\, a_k^j \end{cases}$ (5.12)

Reconstruction: $\quad a_k^j = \sum_{l \in Z} p_{k-2l}a_l^{j-1} + \sum_{l \in Z}(-1)^k \overline{p_{1-k+2l}}\, b_l^{j-1}$ (5.13)

Thus ends our "'Tale of Two Bases." We use the basis $\{\phi_{jk}\}_{k \in Z}$ to view the signal as a whole, and we use the basis $\{\phi_{j-1,k}\}_{k \in Z} \cup \{\psi_{j-1,k}\}_{k \in Z}$ to view the smoothed and oscillatory parts of the signal.

5.1.5   Summary

Let us summarize the main results concerning a multiresolution analysis $\{V_j\}_{j \in Z}$ with scaling function ϕ.

Scaling Function

Basis for V_j: $\{\phi_{jk} = 2^{j/2}\phi(2^j x - k)\}_{k \in Z}$

Scaling Relation: $\begin{cases} \phi(x) = \sum_{k \in Z} p_k\phi(2x - k) \\ \phi(2^{j-1}x - l) = \sum_{k \in Z} p_{k-2l}\phi(2^j x - k) \\ \phi_{j-1,l} = 2^{-1/2}\sum_{k \in Z} p_{k-2l}\phi_{jk} \end{cases}$

Wavelet Spaces

$$W_j \perp V_j \quad \text{and} \quad V_{j+1} = V_j \oplus W_j$$
$$V_{j+1} = \cdots W_{j-2} \oplus W_{j-1} \oplus W_j$$
$$L^2(R) = \cdots W_{j-2} \oplus W_{j-1} \oplus W_j \oplus W_{j+1} \cdots$$

Wavelet

$$\begin{cases} \psi(x) = \sum_{k \in Z} (-1)^k \overline{p_{1-k}} \phi(2x - k) \\[2mm] \psi(2^{j-1}x - l) = \sum_{k \in Z} (-1)^k \overline{p_{1-k+2l}} \phi(2^j x - k) \\[2mm] \psi_{jl} = 2^{-1/2} \sum_{k \in Z} (-1)^k \overline{p_{1-k+2l}} \phi_{j+1,k} \end{cases}$$

Orthonormal Bases

W_j	V_j	$L^2(R)$
$\{\psi_{j-1,k}\}_{k \in Z}$	$\{\phi_{j-1,k}\}_{k \in Z} \cup \{\psi_{j-1,k}\}_{k \in Z}$	$\{\psi_{j,k}\}_{j,k \in Z}$
	$\{\phi_{jk}\}_{k \in Z}$	

Decomposition Formulas

Orthonormal Form $\begin{cases} \langle f, \phi_{j-1,l} \rangle = 2^{-1/2} \sum_{k \in Z} \overline{p_{k-2l}} \langle f, \phi_{jk} \rangle \\[2mm] \langle f, \psi_{j-1,l} \rangle = 2^{-1/2} \sum_{k \in Z} (-1)^k p_{1-k+2l} \langle f, \phi_{jk} \rangle \end{cases}$

Orthogonal Form $\begin{cases} a_l^{j-1} = 2^{-1} \sum_{k \in Z} \overline{p_{k-2l}} a_k^j \\[2mm] b_l^{j-1} = 2^{-1} \sum_{k \in Z} (-1)^k p_{1-k+2l} \, a_k^j \end{cases}$

where

$$f = \sum_{k \in Z} a_k^j \phi(2^j x - k) = \sum_{k \in Z} a_k^{j-1} \phi(2^{j-1}x - k) + b_k^{j-1} \psi(2^{j-1}x - k)$$

Reconstruction Formulas

Orthonormal Form $\begin{cases} \langle f, \phi_{jk} \rangle = 2^{-1/2} \sum_{l \in Z} p_{k-2l} \langle f, \phi_{j-1,l} \rangle \\[2mm] \qquad\qquad + 2^{-1/2} \sum_{l \in Z} (-1)^k \overline{p_{1-k+2l}} \langle f, \psi_{j-1,l} \rangle \end{cases}$

Orthogonal Form $\quad a_k^j = \sum_{l \in Z} p_{k-2l} a_l^{j-1} + \sum_{l \in Z} (-1)^k \overline{p_{1-k+2l}} \, b_l^{j-1}$

5.2 Implementing Decomposition and Reconstruction

In this section, we describe decomposition and reconstruction algorithms associated with a multiresolution analysis. These algorithms are analogous to those presented in the section on Haar wavelets. In later sections, the algorithms developed here will be used in conjunction with a multiresolution analysis involving a continuous scaling function and a continuous wavelet to provide accurate decomposition and reconstruction of signals.

5.2.1   The Decomposition Algorithm

In order to do signal processing, such as filtering or data compression, an efficient algorithm is needed to decompose a signal into parts that contain information about the signal's oscillatory behavior. If we use a multiresolution analysis for this purpose, we need to develop an algorithm for breaking up a signal into its wavelet-space (W_j) components, because these components have the information that we want.

There are three major steps in decomposing a signal f: initialization, iteration, and termination.

Initialization This step involves two parts. First, we have to decide on the approximation space V_j that best fits the information available on f. The sampling rate and our choice of multiresolution analysis determine which V_j to use. Second, we need to choose $f_j \in V_j$ to best fit f itself.

The *best* approximation to f from V_j, in the sense of energy, is $P_j f$, the orthogonal projection of f onto V_j (see Definition 0.19); since $2^{j/2}\phi(2^j x - k)$ is orthonormal, Theorem 0.21 implies

$$P_j f(x) = \sum_{k \in Z} a_k^j \phi(2^j x - k)\,, \text{ where } a_k^j = 2^j \int_{-\infty}^{\infty} f(x)\overline{\phi(2^j x - k)}\,dx\,. \quad (5.14)$$

The information from the *sampled* signal is usually not enough to determine the coefficients a_k^j exactly, and so we have to approximate them using the quadrature rule given in the next theorem.

THEOREM 5.12   *Let $\{V_j, j \in Z\}$ be a given multiresolution analysis with a scaling function ϕ that is compactly supported. If $f \in L^2(R)$ is continuous, then, for j sufficiently large,*

$$a_k^j = 2^j \int_{-\infty}^{\infty} f(x)\overline{\phi(2^j x - k)}\,dx \approx m f(k/2^j),$$

where $m = \int \overline{\phi(x)}\,dx$.

Proof   Since ϕ has compact support, the set where ϕ is nonzero is contained in an interval of the form $\{|t| \le M\}$. (In many cases, M is not large; in fact in the case of the Haar function, $M = 1$, and for the simpler Daubechies wavelets, M is on the order of 5.) Thus, the interval of integration for a_k^j in (5.14) is $\{x;\ |2^j x - k| \le M\}$. By changing variables with $t = 2^j x - k$, we obtain

$$a_k^j = \int_{-M}^{M} f(2^{-j}t + 2^{-j}k)\overline{\phi(t)}\,dt.$$

When j is large, $2^{-j}t + 2^{-j}k \approx 2^{-j}k$ for $t \in [-M, M]$. Therefore, $f(2^{-j}t + 2^{-j}k) \approx f(2^{-j}k)$ for all $t \in [-M, M]$ because f is *uniformly* continuous on any

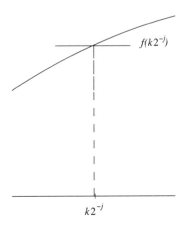

Figure 5 A continuous
function is almost constant
on a small interval

finite interval (this is illustrated in Figure 5.) In particular, we can approximate
the integral on the right by replacing $f(2^{-j}t + 2^{-j}k)$ by $f(2^{-j}k)$ to obtain

$$a_k^j \approx f(k/2^j) \int_{-M}^{M} \overline{\phi(t)}\, dt.$$

Recalling that ϕ is 0 outside of the interval $[-M, M]$, we have that

$$\int_{-M}^{M} \overline{\phi(t)}\, dt = \int_{-\infty}^{\infty} \overline{\phi(t)}\, dt = m,$$

so $a_k^j \approx mf(k/2^j)$.   ◆

The accuracy of this approximation increases with increasing j. In Exercise
12, an estimate is given on how large j must be in order to get an approximation to within a given tolerance. The accuracy may depend on both indicies
j and k as well as the smoothness of f. Since in practice only finitely many
coefficients are dealt with, care should be taken to pick j large enough so that
all of the coefficients are accurately calculated. In addition, the proof of the
theorem can be modified to apply to the case of a piecewise continuous signal
(see Exercise 15) and to the case where ϕ is well localized but not compactly
supported. As mentioned in the proof of Theorem 5.9, $\int \phi \neq 0$ (Exercise 6).
Often, ϕ is constructed so that $\int \phi = 1$ (e.g., Haar and Daubechies), in which
case $a_k^j = f(k/2^j)$. Using the quadrature formula $a_k^j \doteq mf(k/2^j)$, we are also
approximating the projection $P_j f$ by the expansion

$$P_j f(x) \approx f_j(x) = m \sum_{k \in Z} f(k/2^j)\phi(2^j x - k).$$

Note the similarity to the expansions associated with the sampling theorem.

The initialization step discussed here is often simply assumed and used without comment. Strang and Nguyen [20] point this out, and call it the "wavelet crime"!

Iteration   Wavelets shine here! After the initialization step, we have $f \approx f_j \in V_j$. From Theorem 5.11, we can start with f_j and decompose it into a sum of a lower level approximation part, $f_{j-1} \in V_{j-1}$ and a wavelet part, $w_{j-1} \in W_{j-1}$; that is, $f_j = f_{j-1} + w_{j-1}$. We then repeat the process with f_{j-1}, then with f_{j-2}, and so on. This is illustrated in the following diagram, where we have stopped the decomposition at level 0.

$$f \approx f_j \longrightarrow f_{j-1} \longrightarrow f_{j-2} \longrightarrow \cdots \longrightarrow f_1 \longrightarrow f_0$$
$$\searrow w_{j-1} \qquad \searrow w_{j-2} \qquad \searrow \cdots \qquad \searrow w_1 \qquad \searrow w_0$$

To carry out the decomposition, we work with the approximation and wavelet coefficients, the a's and b's that we discussed at the end of Section 5.1.4. As we did for the Haar wavelets, we want to put the decomposition equation (5.12) language of discrete filters and simple operators acting on coefficients.

Recall that the convolution of two sequences $x = (\ldots x_{-1}, x_0, x_1, \ldots)$ and $y = (\ldots y_{-1}, y_0, y_1, \ldots)$ is defined (see Definition 3.9) by

$$(x * y)_l = \sum_{k \in Z} x_k y_{l-k}.$$

Let h and ℓ be the sequences

$$h_k := \frac{1}{2}(-1)^k p_{k+1} \tag{5.15}$$

$$\ell_k := \frac{1}{2}\overline{p_{-k}}. \tag{5.16}$$

Define the two discrete filters (convolution operators) H and L via $H(x) = h * x$ and $L(x) = \ell * x$. Take $x = a^j$ and note that $L(a^j)_l = \frac{1}{2}\sum_{k \in Z} \overline{p_{k-l}} a_k^j$. Comparing this with equation (5.12), we see that $a_l^{j-1} = L(a^j)_{2l}$. Similarly, $b_l^{j-1} = H(a^j)_{2l}$. In discussing the Haar decomposition algorithm, we encountered this operation of discarding the odd coefficients in a sequence and called it *downsampling*. A downsampled signal is a sampling of the signal at every other node and thus corresponds to sampling a signal on a grid that is twice as coarse as the original one. We define the downsampling operator D as follows.

DEFINITION 5.13   *If $x = (\ldots, x_{-2}, x_{-1}, x_0, x_1, x_2, \ldots)$ then its downsample is the sequence*

$$Dx = (\ldots, x_{-2}, x_0, x_2, \ldots)$$

or $(Dx)_l = x_{2l}$ for all $l \in Z$.

We can now formulate the iterative step in the algorithm using discrete filters (convolution operators). The two filters that we use, h and ℓ, are called the *decomposition high-pass* and *decomposition low-pass* filters, respectively.

$$\left.\begin{array}{l} \text{Convolution form:} \quad a^{j-1} = D(\ell * a^j) \quad \text{and} \quad b^{j-1} = D(h * a^j) \\[2mm] \text{Operator form:} \quad a^{j-1} = DLa^j \quad \text{and} \quad b^{j-1} = DHa^j \end{array}\right\} \qquad (5.17)$$

In Figure 6, we illustrate the iterative step in the decomposition algorithm. We represent the downsampling operator pictorially by $2\downarrow$.

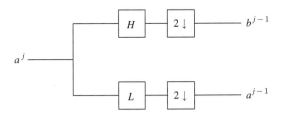

Figure 6 Decomposition diagram for a general multiresolution analysis

It is important to note that the discrete filters and downsampling operator do *not* depend on the level j. Thus storage is minimal and, because convolution is inexpensive computationally, the whole iterative process is fast and efficient.

Termination. There are several criteria for finishing the decomposition. The simplest is that we decompose until we exhaust the finite number of samples we have taken. On the other hand, this may not be necessary. Singularity detection may only require one or two levels. In general, the choice of stopping point greatly depends on what we wish to accomplish.

The end result of the entire decomposition procedure—stopped at $j = 0$, say—is a set of coefficients that includes the approximation coefficients for level 0, (i.e., $\{a_k^0\}$) and the detail (wavelet) coefficients $\{b_k^{j'}\}$ for $j' = 0, \ldots, j-1$.

EXAMPLE 5.14

We consider the signal, f, defined on the unit interval $0 \le x \le 1$ given in Figure 7 (the same one used in Example 4.13). As before, we discretize this signal over 2^8 nodes [so $a_l^8 = f(l/2^8)$]. We now use the decomposition algorithm given in equation (5.17) with the Daubechies wavelets whose p-values are given in (5.4). Plots of the components in V_8, V_7, V_6 and V_4 are given in Figures 8 through 11. Compare these with the corresponding figures for the Haar wavelets (Figures 14, 16, 17, and 18 in Chapter 4). Since the Daubechies wavelets are continuous, they offer better approximations to the original signal than do the Haar wavelets. ∎

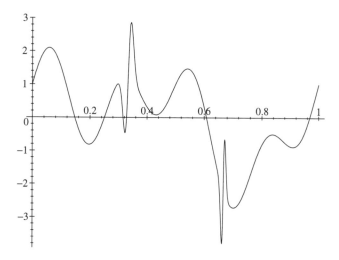

Figure 7 Sample signal

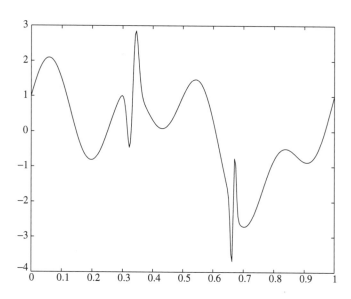

Figure 8 V_8-component with Daubechies

5.2.2 The Reconstruction Algorithm

As mentioned in the section on Haar wavelets, once the signal has been decomposed, some of its $W_{j'}$-components may be modified. If the goal is to filter out noise, then the $W_{j'}$-components of f corresponding to the unwanted frequencies can be thrown out and the resulting signal will have significantly less noise. If the goal is data compression, the $W_{j'}$-components that are small can be thrown out,

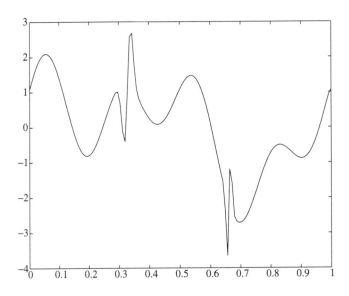

Figure 9 V_7-component with Daubechies

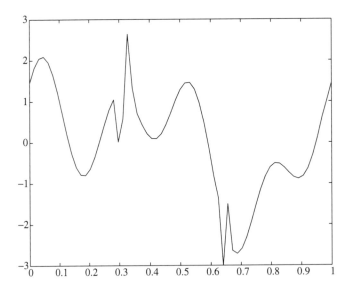

Figure 10 V_6-component with Daubechies

without appreciably changing the signal. Only the significant $W_{j'}$-components (the larger $b_k^{j'}$) need to be transmitted and significant data compression can be achieved. In either case, since the $W_{j'}$-components have been modified, we need a reconstruction algorithm to reassemble the compressed or filtered signal in terms of the basis elements, $\phi(2^j x - l)$, of V_j. The idea is to reconstruct $f_j \approx f$

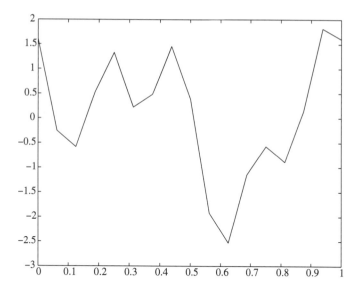

Figure 11 V_4-component with Daubechies

by using $f_{j'} = f_{j'-1} + w_{j'-1}$, starting at $j' = 1$. The scheme is illustrated in the following diagram.

$$f_0 \longrightarrow f_1 \longrightarrow f_2 \longrightarrow \cdots \longrightarrow f_{j-2} \longrightarrow f_{j-1} \longrightarrow f_j \approx f$$
$$w_0 \quad\quad w_1 \quad\quad w_2 \quad\quad \cdots \quad\quad w_{j-2} \quad\quad w_{j-1}$$

We again break up the algorithm into three major steps: initialization, iteration, and termination.

Initialization What we have available is a set of possibly modified coefficients—starting at level 0, say—that includes the approximation coefficients for level 0, $\{a_k^0\}$, and the detail (wavelet) coefficients $\{b_k^{j'}\}$ for $j' = 0, \ldots, j$. These coefficients appear in the expansions

$$f_0(x) = \sum_{k \in Z} a_k^0 \phi(x - k) \in V_0 \tag{5.18}$$

$$w_{j'}(x) = \sum_{k \in Z} b_k^{j'} \psi(2^{j'} x - k) \in W_{j'} \quad \text{for } 0 \leq j' < j. \tag{5.19}$$

Iteration We again formulate this step in terms of discrete filters. Let $\widetilde{h}$ and $\widetilde{\ell}$ be the sequences

$$\widetilde{h}_k := \overline{p_{1-k}}(-1)^k \tag{5.20}$$

$$\widetilde{\ell}_k := p_k. \tag{5.21}$$

Define the two discrete filters (convolution operators) $\widetilde{H}$ and $\widetilde{L}$ via $\widetilde{H}(x) = \widetilde{h} * x$ and $\widetilde{L}(x) = \widetilde{\ell} * x$. The reconstruction formula equation (5.13) gives a_k^j as a sum of two terms, $\sum_{l \in Z} p_{k-2l} a_l^{j-1}$ and $\sum_{l \in Z} (-1)^k \overline{p_{1-k+2l}} b_l^{j-1}$. Using the filter sequences we have just introduced, we can rewrite (5.13) as

$$a_k^j = \sum_{l \in Z} \widetilde{\ell}_{k-2l} a_l^{j-1} + \sum_{l \in Z} \widetilde{h}_{k-2l} b_l^{j-1} \qquad (5.22)$$

This is *almost* a sum of two convolutions; the only difference is that the index for a convolution is $k - l$ instead of $k - 2l$. In other words, (5.22) is a convolution with the odd terms missing (i.e., $\widetilde{\ell}_{k-(2l+1)}$). They can be inserted back by simplify multiplying them by zero. We illustrate this procedure by explicitly writing out the first few terms:

$$a_k^j = \cdots + \widetilde{\ell}_{k+4} a_{-2}^{j-1} + \widetilde{\ell}_{k+3} \cdot 0 + \widetilde{\ell}_{k+2} a_{-1}^{j-1} + \widetilde{\ell}_{k+1} \cdot 0 + \widetilde{\ell}_k a_0^{j-1} + \widetilde{\ell}_{k-1} \cdot 0$$
$$+ \cdots + \text{similar } \widetilde{h}\, b \text{ terms.}$$

In order to put this sum into convolution form, we alter the input sequence a_l^{j-1} by interspersing zeros between its components, making a new sequence that has zeros at all odd entries. Each of the original nonzero terms is given a new, even index by doubling its old index. For example, a_{-1}^{j-1} now has index -2. This procedure is called *upsampling* and is precisely formulated as follows:

DEFINITION 5.15 *Let* $x = (\ldots x_{-2}, x_{-1}, x_0, x_1, x_2, \ldots)$ *be a sequence. We define the* upsampling *operator* U *via*

$$Ux = (\ldots x_{-2}, 0, x_{-1}, 0, x_0, 0, x_1, 0, x_2, 0, \ldots)$$

or

$$(Ux)_k = \begin{cases} 0 & \text{if } k \text{ is odd} \\ x_{k/2} & \text{if } k \text{ is even.} \end{cases}$$

The iterative step (5.22) in the reconstruction algorithm can now be simply formulated in terms of discrete filters (convolution operators).

$$\left. \begin{array}{l} \text{Convolution form:} \quad a^j = \widetilde{\ell} * (U a^{j-1}) + \widetilde{h} * (U b^{j-1}) \\ \text{Operator form:} \quad a^j = \widetilde{L} U a^{j-1} + \widetilde{H} U b^{j-1} \end{array} \right\} \qquad (5.23)$$

We remark that $\widetilde{h}$ and $\widetilde{\ell}$ are called the reconstruction high-pass and low-pass filters, respectively. As was the case in the decomposition algorithm, neither of the filters depends on the level. This makes the iteration in the reconstruction step quick and efficient. In Figure 12, we illustrate the iterative step in the reconstruction algorithm. We represent the upsampling operator pictorially by $2\!\uparrow$.

In the case of the Haar wavelets, $p_0 = p_1 = 1$. This algorithm reduces to the Haar reconstruction algorithm given in Theorem 4.14.

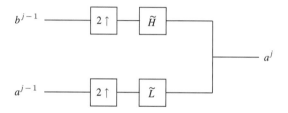

Figure 12 Reconstruction diagram for a
general multiresolution analysis

Termination The decomposition and reconstruction algorithms use the
scaling coefficients, p_k, but not the actual formulas for $\phi(x)$ or $\psi(x)$. To plot
the reconstructed signal $f(x) = \sum_l a_l^j \phi(2^j x - l)$, we can approximate the value of
f at $x = l/2^j$ by a_l^j in view of Theorem 5.12 (provided $\int \phi = 1$). So the formulas
for ϕ and ψ do not even enter into the plot of the reconstructed signal. This is
fortunate since the computation of ϕ and ψ for Daubechies example is rather
complicated (see Sections 5.3.4 and 6.4). However, ϕ and ψ play important
background roles since the orthogonality properties of ϕ and ψ are crucial to
the success of the decomposition and reconstruction algorithms.

5.2.3 Processing a Signal

The steps in processing a signal are essentially the same as the ones discussed
in connection with the Haar system in the previous chapter.

1. *Sample.* This is really a preprocessing step. If the signal is continuous,[1]
   then it must be sampled at a rate sufficient to capture its essential details.
   The sampling rate varies, depending on a variety of factors. For instance,
   to sample with full fidelity a performance of Beethoven's Ninth Symphony,
   we have to capture frequencies up to the audible limit of human hearing,
   roughly 20 kHz. This is the Nyquist frequency. A good rule of thumb
   frequently used in sampling signals is to use a rate that is *double* the
   Nyquist frequency, or 40 kHz in this case.

2. *Decompose.* Once the signal has been sampled, then we use Theorem 5.12
   to approximate the highest-level approximation coefficients. We then iterate equation (5.17) until an appropriate level is reached (such as $j = 0$).
   The output of this step is all levels of wavelet coefficients (details) and the
   lowest-level approximation coefficients. This set of coefficients is the one
   that will be worked with in the next step.

3. *Process the signal.* At this stage, the signal can be compressed by discarding insignificant coefficients, or it can be filtered or denoised in other

---

[1]Continuous signals are frequently called *analog* signals. The term comes from the output
of an analog computer being *continuous*!

ways. The output may be stored or immediately reconstructed to recover the processed version of the signal. In some cases, such as singularity detection, the signal is of no further interest and is discarded.

4. *Reconstruct.* The reconstruction algorithm (5.22) is invoked here. The output of that algorithm is really the set of top-level approximation coefficients. Theorem 5.12 is then invoked to conclude that the processed signal is approximately equal to the top-level reconstructed coefficients.

EXAMPLE 5.16

We apply the decomposition and reconstruction algorithms with Daubechies wavelets to the signal, f, over the unit interval given in Figure 13 (this is the same signal used in Example 4.15).

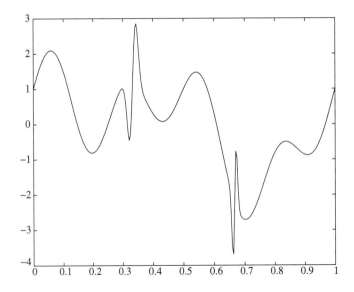

Figure 13 Sample signal

As before, we discretize this signal over 2^8 nodes (so $a_k^8 = f(k/2^8)$ for $0 \leq k \leq 2^8 - 1$) and then decompose this signal with (5.17) to obtain

$$f = f_0 + w_0 + w_1 + w_2 + \cdots + w_7$$

with

$$f_0(x) = a_0^0 \phi(x) \in V_0 \quad \text{and} \quad w_{j'}(x) = \sum_{k \in Z} b_k^{j'} \psi(2^{j'} x - k) \in W_{j'}$$

where ϕ and ψ are the Daubechies scaling function and wavelet (which we will construct shortly). We then compress and reconstruct the signal using (5.22), and we replot the new a_k^8 as the approximate value of the reconstructed signal at $x = k/2^8$. Figures 14 and 15 illustrate 80% and 90% compressions using Daubechies wavelets (by setting the smaller 80% and 90% of the $|b_k^{j'}|$ coefficients equal to zero, respectively). Compare these compressed signals with Figures 20 and 21 in Chapter 4, which illustrate 80% and 90% compression with the Haar wavelets. The relative errors for 80% and 90% compression are

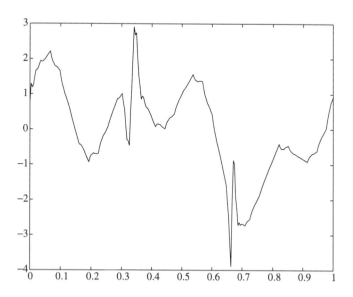

Figure 14 Eighty percent compression with Daubechies

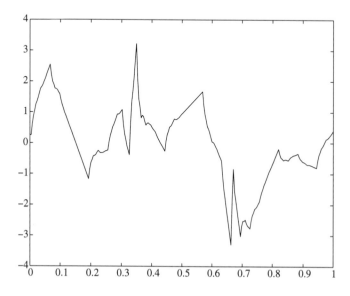

Figure 15 Ninety percent compression with Daubechies

0.0553 and 0.1794, respectively. For comparison, the relative errors for the same compression rates with Haar wavelets are 0.0895 and 0.1838, respectively (see Example 4.15). Using the FFT, for the same compression rates, the relative errors are 0.1228 and 0.1885, respectively.

Do not get the impression from this example that wavelets are better than the FFT for data compression. The signal in this example has two isolated spikes, which wavelets are designed to handle better than the FFT. Signals without spikes or rapid changes or that have an almost periodic nature can be handled quite well with the FFT (perhaps even better than wavelets).   ■

5.3   Fourier Transform Criteria

The examples of multiresolution analyses discussed in the previous sections— Haar, linear splines, and Shannon—all are constructed by *starting* with a given set of sampling spaces. All of these have limitations. For example, as mentioned earlier, the decomposition algorithm based on the Haar scaling function does not provide an accurate approximation to most smooth signals, because the Haar scaling function and associated wavelet are discontinuous. The Shannon multiresolution analysis is smooth enough, but the high- and low-pass filters used in decomposition and reconstruction have infinite length impulse responses (i.e., an infinite number of scaling coefficients). Moreover, these responses are sequences that have slow decay for large index n. The linear splines are better in this regard. Even so, the impulse responses involved are still infinite. In Chapter 6, we construct a multiresolution analysis that has a continuous scaling function; finite impulse responses for the high- and low-pass filters are used in its decomposition and reconstruction algorithms. Instead of starting with the sampling spaces, we construct the scaling function directly using Fourier transform methods. In this section, we state and derive the properties that must be satisfied by a scaling function, and then translate these properties into language involving the Fourier transform.

5.3.1   The Scaling Function

Recall that a multiresolution analysis is, by definition, a collection of subspaces $\{V_j, j \in Z\}$ of $L^2(R)$, where each V_j is the span of $\{\phi(2^j x - k), k \in Z\}$, where ϕ must be constructed so that the collection $\{V_j, j \in Z\}$ satisfies the properties listed in Definition 5.1. The following theorem provides an equivalent formulation of these properties in terms of the function ϕ.

THEOREM 5.17 *Suppose ϕ is a continuous function with compact support satisfying the orthonormality condition: $\int \phi(x - k)\phi(x - l)\,dx = \delta_{kl}$. Let V_j be the span of $\{\phi(2^j x - k); k \in Z\}$. Then the following hold.*

- *The spaces V_j satisfy the separation condition (i.e., $\cap V_j = \{0\}$).*

- *If the following additional conditions are satisfied by ϕ*

1. Normalization: $\int \phi(x)\, dx = 1$

2. Scaling: $\phi(x) = \sum_k p_k \phi(2x - k)$ *for some finite number of constants* p_k

then the associated V_j *satisfy the density condition:* $\cup V_j = L^2(R)$, *or, in other words, any function in* $L^2(R)$ *can be approximated by functions in* V_j *for large enough* j.

In particular, if the function ϕ *is continuous with compact support and satisfies the normalization, scaling, and orthonormality conditions just listed, then the collection of spaces* $\{V_j, j \in Z\}$ *forms a multiresolution analysis.*

Proof Here, we give a nontechnical explanation of the central ideas of the first part of the theorem. Rigorous proofs of the first and second parts are contained in Appendix A.

By definition, a signal, f, belonging to V_{-j} is a linear combination of

$$\{\phi(x/2^j - k), \quad k = \cdots -2, -1, 0, 1, 2 \cdots\}.$$

As $j > 0$ gets large, the set where $\phi(x/2^j - k)$ is nonzero gets larger. For example, if Figure 16 represents the graph of ϕ, then the graph of $\phi(x/2^2)$ is given in Figure 17.

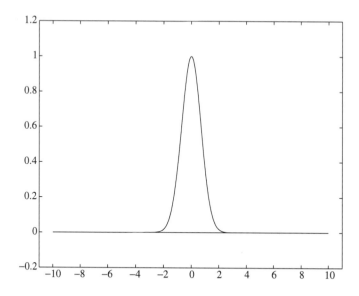

Figure 16 Graph of ϕ

On the other hand, as a member of $L^2(R)$, f has finite energy—that is, $\int |f(t)|^2\, dt$ is finite. As the graphs in Figures 16 and 17 show, the energy of a signal increases ($\int |f(t)|^2\, dt \to \infty$) as the size of the set where the signal is nonzero grows large. Therefore, a *nonzero* signal that belongs to all the V_{-j}, as

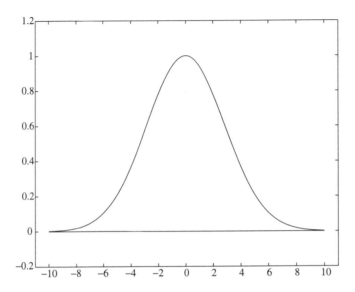

Figure 17 Graph of $\phi(x/4)$

$j \to \infty$, must have infinite energy. We conclude that the only signal belonging
to all the V_j must be identically zero.   ♦

5.3.2   Orthogonality via the Fourier Transform

How is a scaling function ϕ constructed? To construct a scaling function, we
must first find coefficients p_k that satisfy Theorem 5.9. A direct construction
of the p_k is difficult. Instead, we use the Fourier transform, which will more
clearly motivate some of the steps in the construction of the p_k and hence of
ϕ. At first, we assume that a continuous scaling function exists and see what
properties must be satisfied by the associated scaling coefficients p_k. Then
we reverse the process and show that if p_k satisfies these properties, then an
associated scaling function can be constructed. In the next section, an example
of a set of coefficients, p_k, is constructed that satisfies these properties, and this
will lead to the construction of a particular continuous scaling function (due to
Daubechies).

Recall that the definition of the Fourier transform of a function f is given
by

$$\hat{f}(\xi) = \frac{1}{\sqrt{2\pi}} \int_{-\infty}^{\infty} f(x)e^{-ix\xi}\, dx.$$

Since $\hat{\phi}(0) = \frac{1}{\sqrt{2\pi}} \int \phi(x)\, dx$, the normalization condition ($\int \phi = 1$) becomes

$$\hat{\phi}(0) = \frac{1}{\sqrt{2\pi}}.$$

The translations of the orthonormality conditions on ϕ and ψ into the language of the Fourier transform are given in the next theorem.

THEOREM 5.18 *A function ϕ satisfies the orthonormality condition if and only if*

$$2\pi \sum_{k\in Z} |\hat{\phi}(\xi + 2\pi k)|^2 = 1 \quad \text{for all } \xi \in R.$$

In addition, a function $\psi(x)$ is orthogonal to $\phi(x - l)$ for all $l \in Z$ if and only if

$$\sum_{k\in Z} \hat{\phi}(\xi + 2\pi k)\overline{\hat{\psi}(\xi + 2\pi k)} = 0 \quad \text{for all } \xi \in R.$$

Proof We will prove the first part. The proof of the second part is similar. The orthonormality condition can be stated as

$$\int \phi(x - k)\overline{\phi(x - l)}\, dx = \delta_{kl} = \begin{cases} 1, & \text{if } k = l \\ 0, & \text{if } k \neq l. \end{cases}$$

By replacing $x - k$ by x in the preceding integral and relabeling $n = l - k$, the orthonormality condition can be restated as

$$\int \phi(x)\overline{\phi(x - n)}\, dx = \delta_{0n}. \tag{5.24}$$

Recall that Plancherel's identity for the Fourier transform (Theorem 2.12) states

$$\int_{-\infty}^{\infty} f(x)\overline{g(x)}\, dx = \int_{-\infty}^{\infty} \hat{f}(\xi)\overline{\hat{g}(\xi)}\, d\xi \quad \text{for } f,\, g \in L^2.$$

We apply this identity with $f(x) = \phi(x)$ and $g(x) = \phi(x - n)$. By the sixth part of Theorem 2.6 (with $a = n$ and $\lambda = \xi$), $\widehat{\phi(x - n)}(\xi) = \hat{\phi}(\xi)e^{-in\xi}$. So the orthonormality condition (5.24) can be restated as

$$\int_{-\infty}^{\infty} \hat{\phi}(\xi)\overline{\hat{\phi}(\xi)e^{-in\xi}}\, d\xi = \delta_{0n} \quad \text{or} \quad \int_{-\infty}^{\infty} |\hat{\phi}(\xi)|^2 e^{in\xi}\, d\xi = \delta_{0n}.$$

By dividing the real line into the intervals $I_j = [2\pi j, 2\pi(j + 1)]$ for $j \in Z$, this equation can be written as

$$\sum_{j\in Z} \int_{2\pi j}^{2\pi(j+1)} |\hat{\phi}(\xi)|^2 e^{in\xi}\, d\xi = \delta_{0n}.$$

Now replace ξ by $\xi + 2\pi j$. The limits of integration change to 0 and 2π:

$$\int_0^{2\pi} \sum_{j\in Z} |\hat{\phi}(\xi + 2\pi j)|^2 e^{in(\xi + 2\pi j)}\, d\xi = \delta_{0n}.$$

Since $e^{2\pi i j} = 1$, for $j \in Z$, this equation becomes

$$\int_0^{2\pi} \sum_{j \in Z} |\hat{\phi}(\xi + 2\pi j)|^2 e^{in(\xi)} \, d\xi = \delta_{0n}. \tag{5.25}$$

Let

$$F(\xi) = 2\pi \sum_{j \in Z} |\hat{\phi}(\xi + 2\pi j)|^2.$$

The orthonormality condition (5.25) becomes

$$(1/2\pi) \int_0^{2\pi} F(\xi) e^{in\xi} \, d\xi = \delta_{0n}. \tag{5.26}$$

The function F is 2π-periodic because

$$F(\xi + 2\pi) = 2\pi \sum_{j \in Z} |\hat{\phi}(\xi + 2\pi(j+1))|^2$$

$$= 2\pi \sum_{j'} |\hat{\phi}(\xi + 2\pi j')|^2 \quad \text{(replace } j+1 \text{ by } j' \text{)}$$

$$= F(\xi).$$

Since F is periodic, it has a Fourier series, $\sum \alpha_n e^{inx}$, where the Fourier coefficients are given by $\alpha_n = (1/2\pi) \int_0^{2\pi} F(\xi) e^{-in\xi} \, d\xi$. Thus, the orthonormality condition, (5.26), is equivalent to $\alpha_{-n} = \delta_{n0}$, which in turn is equivalent to the statement $F(\xi) = 1$. This completes the proof. ♦

An Interesting Identity. As an application, there is an interesting and useful formula that we can derive using this theorem. Recall that the Haar scaling function ϕ equals 1 on the unit interval and zero everywhere else, and so its Fourier transform is

$$\hat{\phi}(\xi) = \frac{1}{\sqrt{2\pi}} \int_0^1 e^{-ix\xi} \, dx$$

$$= \frac{e^{-i\xi} - 1}{-\sqrt{2\pi} i \xi}. \tag{5.27}$$

With a little algebra, we obtain

$$|\hat{\phi}(\xi)|^2 = \frac{1 - \cos \xi}{\pi \xi^2} = \frac{1}{2\pi} \left(\frac{\sin(\xi/2)}{\xi/2} \right)^2.$$

Since $\{\phi(x-k), \ k \in Z\}$ is an orthonormal set of functions, Theorem 5.18 implies

$$\sum_{k \in Z} \frac{\sin^2 \left(\frac{\xi}{2} + \pi k \right)}{\left(\frac{\xi}{2} + \pi k \right)^2} = 1.$$

Since $\sin^2 x$ is π periodic, all the numerators are equal to $\sin^2 \frac{\xi}{2}$. Dividing by this and simplifying results in the formula

$$\csc^2 \frac{\xi}{2} = \sum_{k \in Z} \frac{4}{(\xi + 2\pi k)^2}. \tag{5.28}$$

This formula is well known, and it is usually derived via techniques from complex analysis. We will apply it in Exercise 10 to get a second scaling function for the linear spline multiresolution analysis.

5.3.3 The Scaling Equation via the Fourier Transform

The scaling condition, $\phi(x) = \sum_k p_k \phi(2x - k)$ can also be recast in terms of the Fourier transform.

THEOREM 5.19 *The scaling condition $\phi(x) = \sum_k p_k \phi(2x - k)$ is equivalent to*

$$\hat{\phi}(\xi) = \hat{\phi}(\xi/2)P(e^{-i\xi/2})$$

where the polynomial P is given by

$$P(z) = \frac{1}{2} \sum_{k \in Z} p_k z^k.$$

Proof We take the Fourier transform of both sides of the equation $\phi(x) = \sum_k p_k \phi(2x - k)$ and then use Theorem 2.6 to obtain

$$\hat{\phi}(\xi) = \frac{1}{2} \sum_{k \in Z} \hat{\phi}(\xi/2)p_k e^{-ik(\xi/2)} = \hat{\phi}(\xi/2)P(e^{-i\xi/2})$$

where $P(z) = (1/2) \sum_k p_k z^k$, as claimed.   ♦

Suppose ϕ satisfies the scaling condition. Using Theorem 5.19 in an iterative fashion, we obtain the following

$$\hat{\phi}(\xi) = P(e^{-i\xi/2})\hat{\phi}(\xi/2)$$
$$\hat{\phi}(\xi/2) = P(e^{-i\xi/2^2})\hat{\phi}(\xi/2^2)$$

and so

$$\hat{\phi}(\xi) = P(e^{-i\xi/2})P(e^{-i\xi/2^2})\hat{\phi}(\xi/2^2).$$

Continuing in this manner, we obtain

$$\hat{\phi}(\xi) = P(e^{-i\xi/2}) \cdots P(e^{-i\xi/2^n})\hat{\phi}(\xi/2^n)$$

$$= \left(\prod_{j=1}^{n} P(e^{-i\xi/2^j}) \right) \hat{\phi}(\xi/2^n).$$

For a given scaling function ϕ, the preceding equation holds for each value of n. In the limit, as $n \to \infty$, this equation becomes

$$\hat{\phi}(\xi) = \left(\prod_{j=1}^{\infty} P(e^{-i\xi/2^j}) \right) \hat{\phi}(0).$$

If ϕ satisfies the normalization condition $\int \phi(x)\,dx = 1$, then $\hat{\phi}(0) = 1/\sqrt{2\pi}$ and so

$$\hat{\phi}(\xi) = \frac{1}{\sqrt{2\pi}} \prod_{j=1}^{\infty} P(e^{-i\xi/2^j}). \tag{5.29}$$

This provides a formula for the Fourier transform of the scaling function ϕ in terms of the scaling polynomial P. This formula is of limited practical use for the construction of ϕ since infinite products are difficult to compute. In addition, the inverse Fourier transform would have to be used to recover ϕ from its Fourier transform. Nevertheless, it is useful theoretically, as we see later.

Given a multiresolution analysis generated by a function ϕ satisfying the scaling condition $\phi(x) = \sum_k p_k \phi(2x - k)$, the associated wavelet, ψ, satisfies the following equation (see Theorem 5.10):

$$\psi(x) = \sum_{k \in Z} (-1)^k \bar{p}_{1-k} \phi(2x - k).$$

Let

$$Q(z) = -z\overline{P(-z)}.$$

For $|z| = 1$, $Q(z) = (1/2) \sum_k (-1)^k \bar{p}_{1-k} z^k$ (see Exercise 13). The same arguments given in the proof of Theorem 5.19 show that

$$\hat{\psi}(\xi) = \hat{\phi}(\xi/2) Q(e^{-i\xi/2}). \tag{5.30}$$

The previous two theorems can be combined to give the following necessary condition on the polynomial $P(z)$ for the existence of a multiresolution analysis.

THEOREM 5.20 *Suppose the function ϕ satisfies the orthonormality condition, $\int \phi(x-k)\phi(x-l)\,dx = \delta_{kl}$, and the scaling condition, $\phi(x) = \sum_k p_k \phi(2x - k)$. Then the polynomial $P(z) = \sum_k p_k z^k$ satisfies the following equation:*

$$|P(z)|^2 + |P(-z)|^2 = 1 \quad \text{for } z \in C \text{ with } |z| = 1$$

or, equivalently,

$$|P(e^{-it})|^2 + |P(e^{-i(t+\pi)})|^2 = 1 \quad \text{for } 0 \le t \le 2\pi.$$

Proof In view of Theorem 5.18, if ϕ satisfies the orthonormality condition, then

$$\sum_{k \in Z} |\hat{\phi}(\xi + 2\pi k)|^2 = 1/2\pi \quad \text{for all } \xi \in R. \tag{5.31}$$

If ϕ satisfies the scaling condition, then (Theorem 5.19)

$$\hat{\phi}(\xi) = \hat{\phi}(\xi/2)P(e^{-i\xi/2}). \tag{5.32}$$

Dividing the sum in (5.31) into even and odd indices and using (5.32), we have

$$\frac{1}{2\pi} = \sum_{k \in Z} |\hat{\phi}(\xi + 2\pi k)|^2$$

$$= \sum_{l \in Z} |\hat{\phi}(\xi + (2l)2\pi)|^2 + \sum_{l \in Z} |\hat{\phi}(\xi + (2l + 1)2\pi)|^2$$

$$= \sum_{l \in Z} \left(|P(e^{-i(\xi/2+2l\pi)})|^2 |\hat{\phi}(\xi/2 + (2l)\pi)|^2 + \right.$$

$$\left. |P(e^{-i(\xi/2+(2l+1)\pi)})|^2 |\hat{\phi}(\xi/2 + \pi(2l + 1))|^2 \right)$$

$$= |P(e^{-i\xi/2})|^2 \sum_{l \in Z} |\hat{\phi}(\xi/2 + 2\pi l)|^2 +$$

$$|P(-e^{-i\xi/2})|^2 \sum_{l \in Z} |\hat{\phi}((\xi/2 + \pi) + 2\pi l)|^2$$

where the last equation uses $e^{-2l\pi i} = 1$ and $e^{-\pi i} = -1$. The two sums appearing on the right are both $1/2\pi$ (using Theorem 5.18 with ξ relaced by $\xi/2$ and $\xi/2 + \pi$). Therefore, the previous equations reduce to

$$1 = |P(e^{-i\xi/2})|^2 + |P(-e^{-i\xi/2})|^2.$$

Since this equation holds for all $\xi \in R$, we conclude that $|P(z)|^2 + |P(-z)|^2 = 1$ for all complex numbers z with $|z| = 1$, as desired. ♦

The following theorem is the analogue of Theorem 5.20 for the function ψ and its associated scaling polynomial Q. Its proof is similar to the proof of the previous theorem and will be left to the reader (see Exercise 14).

THEOREM 5.21 *Suppose the function ϕ satisfies the orthonormality condition, $\int \phi(x-k)\phi(x-l)\,dx = \delta_{kl}$, and the scaling condition, $\phi(x) = \sum_k p_k \phi(2x - k)$. Suppose $\psi(x) = \sum_k q_k \phi(2x - k)$. Let $Q(z) = \sum_k q_k z^k$. Then the following two statements are equivalent:*

- *$\int \psi(x - k)\phi(x - l)\,dx = 0$ for all $k, l \in Z$*

- $P(z)\overline{Q(z)} + P(-z)\overline{Q(-z)} = 0$ *for all* $|z| = 1$

Another way to state Theorems 5.20 and 5.21 is that the matrix

$$M = \begin{pmatrix} P(z) & P(-z) \\ Q(z) & Q(-z) \end{pmatrix}$$

associated to an orthonormal scaling function and orthonormal wavelet must be unitary (that is, $MM^* = I$).

EXAMPLE 5.22

Here, we examine the special case of the Haar scaling function that is one on the unit interval and zero everywhere else. For this example, $p_0 = p_1 = 1$ (all other $p_k = 0$) and so

$$P(z) = (1 + z)/2.$$

We check the statements of Theorems 5.19 and 5.20 for this example. Using the formula for the Fourier transform of the Haar scaling function given in equation (5.27), we obtain

$$P(e^{-i\xi/2})\hat{\phi}(\xi/2) = \frac{1}{2}(1 + e^{-i\xi/2})\left(\frac{e^{-i\xi/2} - 1}{-i\sqrt{2\pi}\xi/2}\right) = \frac{e^{-i\xi} - 1}{-\sqrt{2\pi}i\xi}.$$

So $\hat{\phi}(\xi) = P(e^{-i\xi/2})\hat{\phi}(\xi/2)$, as stated in Theorem 5.19.

Also note that

$$\begin{aligned}
|P(z)|^2 + |P(-z)|^2 &= \frac{|1 + z|^2}{4} + \frac{|1 - z|^2}{4} \\
&= \frac{1 + 2\text{Re}\{z\} + |z|^2}{4} + \frac{1 - 2\text{Re}\{z\} + |z|^2}{4} \\
&= 1 \quad \text{for } z \in C \text{ with } |z| = 1
\end{aligned}$$

in agreement with Theorem 5.20.

In a similar manner, the Fourier transform of the Haar wavelet is

$$\hat{\psi}(\xi) = \frac{(e^{-i\xi/2} - 1)^2}{i\sqrt{2\pi}\xi}.$$

Its scaling polynomial is $Q(z) = (1 - z)/2$ and so

$$\hat{\phi}(\xi/2)Q(e^{-i\xi/2}) = \frac{(e^{-i\xi/2} - 1)}{-i\sqrt{2\pi}\xi/2}\frac{(1 - e^{-i\xi/2})}{2} = \hat{\psi}(\xi)$$

in agreement with equation (5.30).   ∎

5.3.4   Iterative Procedure for Constructing the Scaling Function

The previous theorem states that if ϕ exists, its scaling polynomial P must satisfy the equation $|P(z)|^2 + |P(-z)|^2 = 1$ for $|z| = 1$. So one strategy for constructing a scaling function ϕ is to construct a polynomial P that satisfies this equation and then construct a function ϕ so that it satisfies the scaling equation $\phi(x) = \sum_k p_k \phi(2x - k)$.

Let us assume that P has been constructed to satisfy Theorem 5.20 and that $P(1) = 1$. An example of such a P will be given in the next section. The strategy for constructing the scaling function ϕ associated with P is given by the following iterative process. Let the Haar scaling function be denoted by

$$\phi_0(x) = \begin{cases} 1, & \text{if } 0 \leq x < 1 \\ 0, & \text{otherwise.} \end{cases}$$

The Haar scaling function already satisfies the orthonormality property. Then define

$$\phi_1(x) = \sum_{k \in Z} p_k \phi_0(2x - k). \tag{5.33}$$

In general, define ϕ_n in terms of ϕ_{n-1} by

$$\phi_n(x) = \sum_{k \in Z} p_k \phi_{n-1}(2x - k). \tag{5.34}$$

In the next theorem, we show that the ϕ_n converge, as $n \to \infty$, to a function denoted by ϕ. By taking limits of the preceding equation as $n \to \infty$, we obtain

$$\phi(x) = \sum_{k \in Z} p_k \phi(2x - k)$$

and so ϕ satisfies the desired scaling condition. Since ϕ_0 satisfies the orthonormality condition, there is some hope that ϕ_1 and $\phi_2, \phi_3 \ldots$ and eventually ϕ will also satisfy the orthonormality condition. This procedure turns out to work, under certain additional assumptions on P.

THEOREM 5.23   *Suppose $P(z) = (1/2) \sum_k p_k z^k$ is a polynomial that satisfies the following conditions:*

- *$P(1) = 1$*

- *$|P(z)|^2 + |P(-z)|^2 = 1$ for $|z| = 1$*

- *$|P(e^{it})| > 0$ for $|t| \leq \pi/2$*

Let $\phi_0(x)$ be the Haar scaling function and let $\phi_n(x) = \sum_k p_k \phi_{n-1}(2x - k)$ for $n \geq 1$. Then the sequence ϕ_n converges pointwise and in L^2 to a function ϕ, which satisfies the orthonormality condition

$$\int_{-\infty}^{\infty} \phi(x - n)\phi(x - m)\, dx = \delta_{nm}$$

and which satisfies the scaling equation, $\phi(x) = \sum_k p_k \phi(2x - k)$.

Proof A formal proof of the convergence part of the theorem is technical and is contained in Appendix A. Once convergence has been established, ϕ automatically satisfies the scaling condition, as already mentioned. Here, we give an explanation of why ϕ satisfies the normalization and orthonormality conditions. We first start with ϕ_1, which by (5.33) is

$$\phi_1(x) = \sum_{k \in Z} p_k \phi_0(2x - k).$$

This is equivalent to the following Fourier transform equation:

$$\hat{\phi}_1(\xi) = P(e^{-i\xi/2})\hat{\phi}_0(\xi/2) \tag{5.35}$$

by the same argument as in the proof of Theorem 5.19. Since $\hat{\phi}_0(0) = \frac{1}{\sqrt{2\pi}}$ and $P(1) = 1$, clearly $\hat{\phi}_1(0) = \frac{1}{\sqrt{2\pi}}$ and so ϕ_1 satisfies the normalization condition.

By (5.35), we have

$$\sum_{k \in Z} |\hat{\phi}_1(\xi + 2\pi k)|^2 = \sum_{k \in Z} |P(e^{-i\xi/2 + \pi ki})|^2 |\hat{\phi}_0(\xi/2 + \pi k)|^2.$$

Dividing the sum on the right into even and odd indices and using the same argument as in the proof of Theorem 5.20, we obtain

$$\sum_{k \in Z} |\hat{\phi}_1(\xi + 2\pi k)|^2 = \sum_{l \in Z} |P(e^{-i\xi/2 + 2\pi li})|^2 |\hat{\phi}_0(\xi/2 + 2\pi l)|^2$$

$$+ \sum_{l \in Z} |P(e^{-i\xi/2 + \pi(2l+1)i})|^2 |\hat{\phi}_0(\xi/2 + \pi(2l + 1))|^2$$

$$= |P(e^{-i\xi/2})|^2 \sum_{l \in Z} |\hat{\phi}_0(\xi/2 + 2\pi l)|^2$$

$$+ |P(-e^{-i\xi/2})|^2 \sum_{l \in Z} |\hat{\phi}_0(\xi/2 + \pi + 2\pi l)|^2.$$

Since ϕ_0 satisfies the orthonormality condition, both sums on the right side equal $1/2\pi$ (by Theorem 5.18). Therefore,

$$\sum_{k \in Z} |\hat{\phi}_1(\xi + 2\pi k)|^2 = \frac{1}{2\pi} \left(|P(e^{-i\xi/2})|^2 + |P(-e^{-i\xi/2})|^2 \right)$$

$$= \frac{1}{2\pi} \quad \text{by the assumption on } P.$$

By Theorem 5.18, ϕ_1 satisfies the orthonormality condition. The same argument as before shows that if ϕ_{n-1} satisfies the normalization and orthonormality conditions, then so does ϕ_n. By induction, all the ϕ_n satisfy these conditions. After taking limits, we conclude that ϕ satisfies these conditions as well. ◆

EXAMPLE 5.24

In the case of Haar, $p_0 = p_1 = 1$ and all other $p_k = 0$. In this case, $P(z) = (1 + z)/2$. Clearly $P(1) = 1$, and in Example 5.22 we showed that $|P(z)|^2 + |P(-z)|^2 = 1$ for $|z| = 1$. Also, $|P(e^{it})| = |(1 + e^{it})/2| > 0$ for all $|t| < \pi$. So all the conditions of Theorem 5.23 are satisfied. The iterative algorithm in this theorem starts with the Haar scaling function for ϕ_0. For the Haar scaling coefficients, $p_0 = p_1 = 1$, all the ϕ_k equal ϕ_0. For example, from (5.33),

$$\begin{aligned} \phi_1(x) &= p_0\phi_0(2x) + p_1\phi_0(2x - 1) \\ &= \phi_0(2x) + \phi_0(2x - 1) \\ &= \phi_0(x) \quad \text{from (4.12)}. \end{aligned}$$

Likewise, $\phi_2 = \phi_1 = \phi_0$, and so forth. This is not surprising since the iterative algorithm uses the scaling equation to produce ϕ_k from ϕ_{k-1}. If the Haar scaling equation is used, with the Haar scaling function as ϕ_0, then the iterative algorithm will keep reproducing ϕ_0. ∎

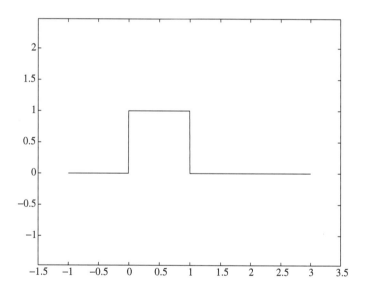

Figure 18 Plot of ϕ_0

EXAMPLE 5.25

Daubechies Example Let $P(z) = (1/2) \sum_k p_k z^k$, where

$$p_0 = \frac{1 + \sqrt{3}}{4} \quad p_1 = \frac{3 + \sqrt{3}}{4} \quad p_2 = \frac{3 - \sqrt{3}}{4} \quad p_3 = \frac{1 - \sqrt{3}}{4}.$$

As established in Exercise 11, $P(z)$ satisfies the conditions in Theorem 5.23. Therefore, the iterative scheme stated in Theorem 5.23 converges to a scaling

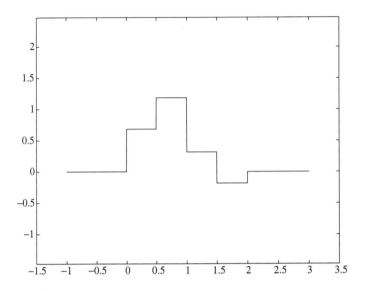

Figure 19 Plot of ϕ_1 for Daubechies

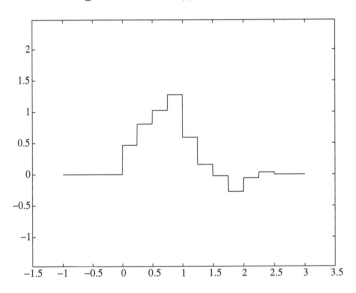

Figure 20 Plot of ϕ_2 for Daubechies

function ϕ whose construction is due to Daubechies. Plots of the first four iterations, ϕ_0, ϕ_1, ϕ_2 and ϕ_4, of this process are given in Figures 18 through 21.

∎

In the next chapter, we motivate the calculation of this choice of $p_0, \ldots, p_3$ and other similar choices of p_k.

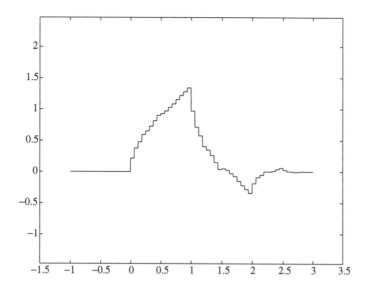

Figure 21 Plot of ϕ_4 for Daubechies

5.4 Exercises

1. For each function g, plot $g(x)$, $g(x+1)$, $2^{1/2}g(2x-3)$, and $\frac{1}{2}g(\frac{1}{4}x-1)$.

   (a) $g(x) = e^{-x^2}$

   (b) $g(x) = \frac{1}{1+x^2}$

   (c) $g(x) = \begin{cases} (1-x^2)^2 & |x| < 1 \\ 0 & |x| \geq 1 \end{cases}$

   (d) $g(x) = \dfrac{\sin(\pi x)}{\pi x}$

2. The *support* of a complex-valued function f is the smallest closed set that contains the set where $f \neq 0$. We denote the support of f by $\mathrm{supp}(f)$. We say that f is *compactly supported* or has *compact support* if $\mathrm{supp}(f)$ is contained in a bounded set. Find the support of the following functions. Which are compactly supported?

   (a) $f(x) := \begin{cases} 1 & \text{if } k < x < k+1 \text{ for } k \text{ odd} \\ 0 & \text{if } k \leq x \leq k+1 \text{ for } k \text{ even} \end{cases}$

   (b) $g(x) := \begin{cases} x(1-x) & \text{if } 0 \leq x \leq 3 \\ 1 & \text{if } x > 3 \\ -1 & \text{if } -5 < x < 0 \\ 0 & \text{if } x \leq -5 \end{cases}$

(c) $h(r, \theta) := \begin{cases} r^2(1 - r^2) & \text{if } 0 \le r < 1 \text{ and } 0 \le \theta < 2\pi \\ 0 & \text{otherwise} \end{cases}$

3. Parseval's equation came up in connection with Fourier series, in Theorem 1.40. It holds under much more general circumstances, however. Let V be a complex inner product space with an orthonormal basis $\{u_k\}_{k=1}^{\infty}$. Show that if $f \in V$ and $g \in V$ have the expansions

$$f = \sum_{k=1}^{\infty} a_k u_k \quad \text{and} \quad g = \sum_{k=1}^{\infty} b_k u_k,$$

then

$$\langle f, g \rangle = \sum_{k=1}^{\infty} a_k \bar{b}_k.$$

This is called *Parseval's equation*. The indexing $k = 1, 2, \dots$ is unimportant. Any index set (including double indexing) can be used just long as the index set is countable.

4. Use Parseval's equation from Exercise 3 to establish the following equations.

   (a) Equation 1 in Theorem 5.9.

   (b) Show that

   $$\langle \psi_{0m}, \psi_{0l} \rangle = \frac{1}{2} \sum_{k \in Z} \overline{p_{1-k+2m}}\, p_{1-k+2l},$$

   where ψ is defined in equation (5.5). *Hint*: Change summation indices in equation (5.5) and use the fact that $\{2^{1/2}\phi(2x - k)\}_{k \in Z}$ is orthonormal.

   (c) Show that

   $$\langle \phi_{0l}, \psi_{0m} \rangle = \frac{1}{2} \sum_{k \in Z} (-1)^k \overline{p_{1-k+2m}}\, p_{k-2l}$$

   where ψ is defined in equation (5.5). *Hint*: Change summation indices in equation (5.5) and use the fact that $\{2^{1/2}\phi(2x - k)\}_{k \in Z}$ is orthonormal.

5. We defined the support of a function in Exercise 2. Show that if the scaling function ϕ in a multiresolution analysis has compact support, then there is only a finite number of nonzero coefficients in the scaling relation from Theorem 5.6.

6. Suppose that $\{V_j : j \in Z\}$ is a multiresolution analysis with scaling function ϕ and that ϕ is continuous and compactly supported.

   (a) Find u_j, the orthogonal projection onto V_j, of the step function

   $$u(x) = \begin{cases} 1 & 0 \le x \le 1 \\ 0 & x < 0 \text{ or } x > 1. \end{cases}$$

(b) If $\int_{-\infty}^{\infty} \phi(x)\,dx = 0$, show that for all j sufficiently large, $\|u - u_j\| \geq \frac{1}{2}$.

(c) Explain why the preceding result implies that $\int_{-\infty}^{\infty} \phi(x)\,dx \neq 0$.

7. In case the p_k's are complex, the quantities E and O, which are defined in connection with the proof of equation (4) in Theorem 5.9, may be complex. Show that even if we allow for this, the equations $|E|^2 + |O|^2 = 2$ and $E + O = 2$ have $E = O = 1$ as their *only* solution.

8. (*Shannon Multiresolution Analysis*). For $j \in Z$, let V_j be the space of all finite energy signals f for which the Fourier transform $\hat{f} = 0$ outside of the interval $[-2^j\pi, 2^j\pi]$—that is, all $f \in L^2(R)$ that are band limited and have $\mathrm{supp}(\hat{f}) \subseteq [-2^j\pi, 2^j\pi]$.

(a) Show that $\{V_j\}_{j\in Z}$ satisfies properties 1–4 in the definition of a multiresolution analysis, Definition 5.1.

(b) Show that $\phi(x) := \mathrm{sinc}(x)$, where sinc is defined in equation (5.1), satisfies property 5 of Definition 5.1, and so it is a scaling function for the V_j's. (*Hint*: Use the sampling theorem 2.23 to show that $\{\phi(x-k)\}$ is an orthonormal basis for V_0.)

(c) Show that ϕ satisfies the scaling relation

$$\phi(x) = \phi(2x) + \sum_{k\in Z} \frac{2(-1)^k}{(2k+1)\pi} \phi(2x - 2k - 1).$$

(d) Find an expansion for the wavelet ψ associated with ϕ.

(e) Find the high- and low-pass decomposition filters, h and ℓ.

(f) Find the high- and low-pass reconstruction filters, $\tilde{h}$ and $\tilde{\ell}$.

9. (*Linear Spline Multiresolution Analysis*). For $j \in Z$, let V_j be the space of all finite energy signals f that are continuous and piecewise linear, with possible corners occurring only at the dyadic points $k/2^j$, $k \in Z$.

(a) Show that $\{V_j\}_{j\in Z}$ satisfies properties 1–4 in the definition of a multiresolution analysis, Definition 5.1.

(b) Let $\varphi(x)$ be the "tent function,"

$$\varphi(x) = \begin{cases} x + 1, & -1 \leq x \leq 0, \\ 1 - x, & 0 < x \leq 1, \\ 0, & |x| > 1. \end{cases}$$

Show that $\{\varphi(x - k)\}_{k\in Z}$ is a (nonorthogonal) basis for V_0. Find the scaling relation for for φ.

10. Let $\varphi(x)$ be the tent function from Exercise 9.

(a) Show that

$$\widehat{\varphi}(\xi) = 2\sqrt{\frac{2}{\pi}} \frac{\sin^2(\xi/2)}{\xi^2}.$$

(b) Show that

$$\sum_{k \in Z} \frac{1}{(\xi + 2\pi k)^4} = \frac{3 - 2\sin^2(\xi/2)}{48 \sin^4(\xi/2)}.$$

[*Hint*: Differentiate both sides of equation (5.28) twice and simplify.]

(c) Verify that the function ϕ defined via its Fourier transform,

$$\hat{\phi}(\xi) := 2\sqrt{\frac{2}{\pi}} \frac{\sin^2(\xi/2)}{\xi^2 \sqrt{1 - \frac{2}{3}\sin^2\frac{\xi}{2}}},$$

is such that $\{\phi(x - k)\}_{k \in Z}$ is an orthonormal set. [*Hint*: Use (b) to show that $\sum_{k \in Z} \left|\widehat{\phi}(\xi + 2\pi k)\right|^2 = \frac{1}{2\pi}$, and then apply Theorem 5.18.]

11. Let $P(z) = (1/2) \sum_{k=0}^3 p_k z^k$, where

$$p_0 = \frac{1 + \sqrt{3}}{4} \quad p_1 = \frac{3 + \sqrt{3}}{4} \quad p_2 = \frac{3 - \sqrt{3}}{4} \quad p_3 = \frac{1 - \sqrt{3}}{4}.$$

Explicitly show that $P(z)$ satisfies the conditions in Theorem 5.23. Use a computer algebra system, such as Maple, to establish the equation $|P(z)|^2 + |P(-z)|^2 = 1$ for $|z| = 1$. Show by a graph of $y = |P(e^{it})|$ that $|P(e^{it})| > 0$ for $|t| \leq \pi/2$.

12. Suppose f is a continuously differentiable function with $|f'(x)| \leq M$ for $0 \leq x < 1$. Use the following outline to approximate f uniformly to within some specified tolerance, ϵ, by a step function in V_n, the space generated by $\phi(2^n x - k)$, $k \in Z$, where ϕ is the Haar scaling function.

(a) For $1 \leq j \leq 2^n$, let $a_j = f(j/2^n)$ and form the step function

$$f_n(x) = \sum_{j \in Z} a_j \phi(2^n x - j).$$

(b) Show that $|f(x) - f_n(x)| \leq \epsilon$ provided that n is greater than $\log_2(M/\epsilon)$.

13. Let $P(z) = \sum_k p_k z^k$ and define

$$Q(z) = -z\overline{P(-z)}.$$

For $|z| = 1$, show that $Q(z) = (1/2) \sum_k (-1)^k \overline{p}_{1-k} z^k$. This exercise establishes a formula for the scaling polynomial for the wavelet, ψ, from the scaling polynomial, P, for the scaling function.

14. Prove Theorem 5.21 using the ideas in the proof of Theorem 5.20.

15. Prove Theorem 5.12 in the case where the signal f is only piecewise continuous by dividing the support of f into subintervals where f is continuous. Care must be used for the nodes $l/2^j$ that lie near the discontinuities, for then the approximation $f(x) \approx f(l/2^j)$ does not necessarily hold. However, these troublesome nodes are fixed in number and the support of $\phi(2^j x - l)$ becomes small as j becomes large. So the L^2 contribution of the terms $\alpha_l \phi(2^j x - l)$ corresponding to the troublesome nodes can be shown to be small as j gets large.

16. Prove the second part of Theorem 5.18 using the proof of the first part as a guide.

17. In Exercise 5, we showed that if ϕ has compact support, then only a finite number of p_k's are nonzero. Use the iterative construction of ϕ in equation (5.34) to show that if only a finite number of p_k's are nonzero, then ϕ has compact support.

18. (*Filter Length and Support.*) Show that if the p_k's in the scaling relation are all zero for $k > N$ and $k < 0$, and if the iterates ϕ_n converge to ϕ, then ϕ has compact support. Find the support of ϕ in terms of N.

Chapter 6

The Daubechies Wavelets

The wavelets that we have looked at so far—Haar, Shannon, and linear spline wavelets—all have major drawbacks. Haar wavelets have compact support but are discontinuous. Shannon wavelets are very smooth but extend throughout the whole real line, and they decay at infinity very slowly. Linear spline wavelets are continuous, but the orthogonal scaling function and associated wavelet, like the Shannon wavelets, have infinite support; they do, however, decay rapidly at infinity.

These wavelets, together with a few others having similar properties, were the only ones available before Ingrid Daubechies discovered the hierarchy of wavelets that are named after her. The simplest of these is just the Haar wavelet, which is the only discontinuous one. The other wavelets in the hierarchy are compactly supported and continuous. Better still, by going up the hierarchy, they become *increasingly* smooth; that is, they can have have a prescribed number of continuous derivatives. The wavelet's smoothness can be chosen to satisfy conditions in a particular application. We now turn to Daubechies's construction of the first wavelet past Haar, ψ_2.

6.1 Daubechies's Construction

Theorem 5.23 lists the three sufficient conditions on the polynomial P that ensures that the iteration scheme just described produces a scaling function. For a given polynomial $P(z)$, let

$$p(\xi) = P(e^{-i\xi}).$$

In terms of the of the function p, the three conditions in the hypothesis of Theorem 5.23 can be stated as

$$p(0) = 1 \tag{6.1}$$

$$|p(\xi)|^2 + |p(\xi + \pi)|^2 = 1 \tag{6.2}$$

$$|p(\xi)| > 0 \quad \text{for } -\pi/2 \le \xi \le \pi/2. \tag{6.3}$$

In this section, we describe one polynomial, due to Daubechies, that satisfies (6.1) through (6.3).

From Example 5.22, the polynomial associated with the Haar scaling function is

$$p_0(\xi) = P(e^{-i\xi}) = \frac{1 + e^{-i\xi}}{2} = e^{-i\xi/2} \cos(\xi/2).$$

This choice of p_0 satisfies (6.1) through (6.3). However, the Haar scaling function is discontinuous. One way to generate a continuous scaling function is to take convolution powers. In fact, the convolution of the Haar scaling function with itself can be shown to equal the following continuous linear spline (see Exercise 5 in Chapter 2):

$$\phi_0 * \phi_0(x) = \begin{cases} 1 - |x - 1| & \text{for } 0 \le x \le 2 \\ 0 & \text{otherwise} \end{cases}$$

(see Figure 1). Now the Fourier transform of a convolution equals the product of the Fourier transforms. In particular, the Fourier transform of $\phi * \phi * \cdots * \phi$ (n times) is $(2\pi)^{n/2}(\hat{\phi})^n$. In view of equation (5.29) for the Fourier transform of the

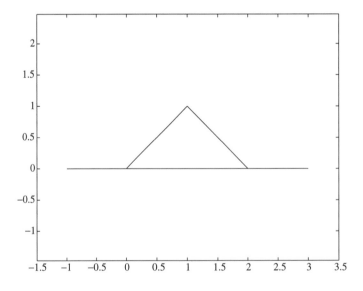

Figure 1 Graph of $\phi_0 * \phi_0$

scaling function, we may be first led to try $p(\xi) = p_0(\xi)^n = e^{-in\xi/2}(\cos \xi/2)^n$ for some suitable power of n as the function that generates a continuous scaling function. However, property (6.2) no longer holds (unless $n = 1$—the Haar case) and so this first attempt fails.

Instead of simply raising p_0 to the nth power, we raise both sides of the identity $\cos^2(\xi/2) + \sin^2(\xi/2) = 1$ to the nth power. With $n = 3$, we obtain

$$1 = \left(\cos^2(\xi/2) + \sin^2(\xi/2)\right)^3 \tag{6.4}$$

or

$$1 = \cos^6(\xi/2) + 3\cos^4(\xi/2)\sin^2(\xi/2) + 3\cos^2(\xi/2)\sin^4(\xi/2) + \sin^6(\xi/2).$$

Using the identities $\cos(u) = \sin(u + \pi/2)$ and $\sin(u) = -\cos(u + \pi/2)$ for the last two terms on the right, we have

$$1 = \cos^6(\xi/2) + 3\cos^4(\xi/2)\sin^2(\xi/2) + 3\sin^2((\xi+\pi)/2)\cos^4((\xi+\pi)/2)$$
$$+ \cos^6((\xi+\pi)/2).$$

If we take

$$|p(\xi)|^2 = \cos^6(\xi/2) + 3\cos^4(\xi/2)\sin^2(\xi/2),$$

then the previous equation becomes

$$1 = |p(\xi)|^2 + |p(\xi + \pi)|^2$$

and property (6.2) is satisfied. Property (6.3) is also satisfied since $\cos(\xi/2) \geq 1/\sqrt{2}$ for $|\xi| \leq \pi/2$. Also note that $|p(0)| = 1$. So all that is left is to identify p itself (we have only defined $|p|$). First, we rewrite the defining equation for $|p|$ as

$$|p(\xi)|^2 = \cos^4(\xi/2)\left(\cos^2(\xi/2) + 3\sin^2(\xi/2)\right)$$
$$= \cos^4(\xi/2)|\cos(\xi/2) + \sqrt{3}i\sin(\xi/2)|^2.$$

By taking square roots of this equation, we choose

$$p(\xi) = \cos^2(\xi/2)\left(\cos(\xi/2) + \sqrt{3}i\sin(\xi/2)\right)\alpha(\xi)$$

where $\alpha(\xi)$ is a complex-valued expression with $|\alpha(\xi)| = 1$ that will be chosen later.

To identify the polynomial P [with $p(\xi) = P(e^{-i\xi})$], we use the identities

$$\cos(\xi/2) = \frac{e^{i\xi/2} + e^{-i\xi/2}}{2} \qquad \sin(\xi/2) = \frac{e^{i\xi/2} - e^{-i\xi/2}}{2i}$$

to obtain

$$p(\xi) = \frac{1}{8}(e^{i\xi} + 2 + e^{-i\xi})\left(e^{i\xi/2} + e^{-i\xi/2} + \sqrt{3}e^{i\xi/2} - \sqrt{3}e^{-i\xi/2}\right)\alpha(\xi).$$

We choose $\alpha(\xi) = e^{-3i\xi/2}$ in order to clear all positive and fractional powers in the exponent. Expanding out and collecting terms, we obtain

$$p(\xi) = \left(\frac{1+\sqrt{3}}{8}\right) + e^{-i\xi}\left(\frac{3+\sqrt{3}}{8}\right) + e^{-2i\xi}\left(\frac{3-\sqrt{3}}{8}\right) + e^{-3i\xi}\left(\frac{1-\sqrt{3}}{8}\right).$$

The equation $p(\xi) = P(e^{-i\xi})$ is therefore satisfied by the polynomial

$$P(z) = \left(\frac{1+\sqrt{3}}{8}\right) + \left(\frac{3+\sqrt{3}}{8}\right)z + \left(\frac{3-\sqrt{3}}{8}\right)z^2 + \left(\frac{1-\sqrt{3}}{8}\right)z^3.$$

Since p satisfies (6.1) through (6.3), P satisfies the hypothesis of Theorem 5.23. Recall that $P(z) = (1/2)\sum_k p_k z^k$. Therefore, for Daubechies's example,

$$p_0 = \frac{1+\sqrt{3}}{4} \qquad p_1 = \frac{3+\sqrt{3}}{4} \qquad p_2 = \frac{3-\sqrt{3}}{4} \qquad p_3 = \frac{1-\sqrt{3}}{4}. \qquad (6.5)$$

The Daubechies scaling function, ϕ, is found by applying the iterative procedure described in Theorem 5.23. Figures 19 through 21 in Chapter 5 illustrate the first four iterations of this procedure. Figure 2 shows the (approximate) graph of ϕ that results from iterating this procedure many times.

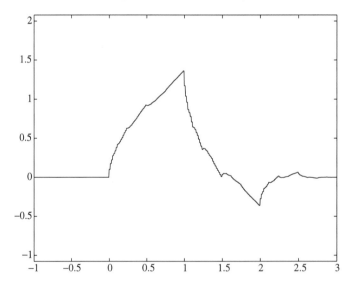

Figure 2 Daubechies scaling function

What about the wavelet ψ that is associated with the Daubechies scaling function ϕ? As shown in Theorem 5.10, once the coefficients, p_k, have been identified and ϕ has been constructed, then the associated wavelet is given by the formula

$$\psi(x) = \sum_{k \in Z} (-1)^k p_{1-k} \phi(2x - k).$$

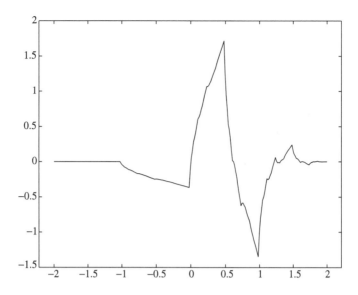

Figure 3 Daubechies wavelet function

Figure 3 shows the (approximate) graph of the associated wavelet function.

Unlike the Haar scaling and wavelet functions, these Daubechies scaling and wavelet functions are continuous. However, they are not differentiable.

6.2   Classification, Moments, and Smoothness

Other smoother scaling functions and wavelets can be obtained by choosing a higher power than $n = 3$ in equation (6.4). Any odd power, $n = 2N - 1$, can be used there. In fact, Daubechies showed that for every N there will be $2N$ nonzero, real scaling coefficients $p_0, \ldots, p_{2N-1}$, resulting in a scaling function and wavelet that are supported on the interval $0 \leq t \leq 2N - 1$. They are chosen so that the corresponding degree $2N - 1$ polynomial $P_N(z) := \frac{1}{2} \sum_{k=0}^{2N-1} p_k z^k$ has the factorization

$$P_N(z) = (z + 1)^N \tilde{P}_N(z) \tag{6.6}$$

where the degree of $\widetilde{P}_N$ is $N - 1$ and $\tilde{P}_N(-1) \neq 0$. This guarantees that the associated wavelets will have precisely N "vanishing moments." We discuss what this means and why it is important later.

Apart from a reversal of the coefficients in P_N (cf. Exercise 1), these coefficients are unique. In the two cases that we have dealt with, $N = 1$ (Haar) and $N = 2$ (Daubechies), the polynomials are $P_1(z) = \frac{1}{2}(1 + z)^1$ and $P_2(z) = \left(\frac{1+\sqrt{3}}{8} + \frac{1-\sqrt{3}}{8} z \right) (1+z)^2$. Both polynomials thus have the factorization in (6.6), with $\widetilde{P}_1(x) = \frac{1}{2}$ and $\widetilde{P}_2(x) = \frac{1+\sqrt{3}}{8} + \frac{1-\sqrt{3}}{8} z$.

The scaling function ϕ_N and wavelet ψ_N generated by P_N have Fourier transforms given in terms of infinite products. Because the coefficients of P_N are real, $\overline{P_N(-z)} = P_N(-\bar{z})$. Thus from equations (5.29) and (5.30), we have that

$$\hat{\phi}_N(\xi) = \frac{1}{\sqrt{2\pi}} \prod_{j=1}^{\infty} P_N(e^{-i\xi/2^j}) \tag{6.7}$$

$$\hat{\psi}_N(\xi) = -e^{-i\xi/2} P_N(-e^{i\xi/2}) \, \hat{\phi}_N(\xi/2). \tag{6.8}$$

Note that $\hat{\psi}_N(0) = 0$, because $P_N(-1) = 0$. If $N > 1$, we also have $\hat{\psi}_N'(0) = 0$, because $P_N'(-1) = 0$. In general, by Exercise 2, we have that

$$\hat{\psi}_N^{(k)}(0) = \begin{cases} 0, & k = 0, \dots, N-1, \\ -N!(-i/2)^N \widetilde{P}_N(-1)/\sqrt{2\pi} \neq 0, & k = N. \end{cases} \tag{6.9}$$

This gives the following result.

PROPOSITION 6.1   *For the Daubechies wavelet ψ_N, we have*

$$\int_{-\infty}^{\infty} x^k \psi_N(x)\, dx = \begin{cases} 0, & k = 0, \dots, N-1, \\ -\left(2^{-N} N!/\sqrt{2\pi}\right) \widetilde{P}_N(-1), & k = N. \end{cases} \tag{6.10}$$

Proof   The proposition follows from equation (6.9) and the Fourier transform property (Theorem 2.6, property 2):

$$\mathcal{F}[t^n f(t)](\lambda) = i^n \frac{d^n}{d\lambda^n} \mathcal{F}[f(t)](\lambda).$$

Just set $f = \psi_N$, $n = k$, and $\lambda = 0$.   ♦

In mechanics, integrals of the form $\int_{-\infty}^{\infty} x^k \rho(x)\, dx$ are called the *moments* of a mass distribution ρ. The term *moment* carries over to an integral of any function against x^k. We can thus rephrase the proposition by saying that ψ_N has its first N moments vanish. This is usually shortened to saying that ψ_N has N vanishing moments; that they are the *first* N is understood.

Daubechies wavelets are classified according to the number of vanishing moments they have. The smoothness of the scaling function and wavelet increase with the number of vanishing moments. The $N = 1$ case is the same as the Haar case; both scaling function and wavelet are discontinuous. The $N = 2$ Daubechies scaling function and wavelet are continuous but certainly do not have smooth derivatives. In the $N = 3$ case, both *are* continuously differentiable. When N is large, the number of continuous derivatives that ϕ_N and ψ_N have is roughly $\frac{N}{5}$. So, to get 10 derivatives, we need to take $N \approx 50$. In the following table, we list approximate scaling coefficients for the Daubechies wavelets, with N going from 1 to 4. Of course, the scaling coefficients given for $N = 2$ are just the decimal approximations for those found in (6.5).

Scaling Coefficient p_k	Number of Vanishing Moments			
	$N = 1$	$N = 2$	$N = 3$	$N = 4$
p_0	1	0.683013	0.470467	0.325803
p_1	1	1.183013	1.141117	1.010946
p_2	0	0.316987	0.650365	0.892200
p_3	0	-0.183013	-0.190934	-0.039575
p_4	0	0	-0.120832	-0.264507
p_5	0	0	0.049817	0.043616
p_6	0	0	0	0.023252
p_7	0	0	0	-0.014987

Why is it useful to have vanishing moments? The short answer is that vanishing moments are a key factor in many wavelet applications—compression, noise removal, singularity detection, for example.

To understand this better, let's look closely at the $N = 2$ case. According to equation (6.10) with $N = 2$, the first two moments ($k = 0, 1$) vanish. The $k = 2$ moment is $-\left(2^{-1}/\sqrt{2\pi}\right) \widetilde{P}_2(-1)$. From our earlier discussion, we see that $\widetilde{P}_2(-1) = \frac{1+\sqrt{3}}{8} - \frac{1-\sqrt{3}}{8} = \frac{\sqrt{3}}{4}$. Thus, for the third ($k = 2$) moment, we have

$$\int_{-\infty}^{\infty} x^2 \psi_2(x)\, dx = -\frac{1}{8}\sqrt{\frac{3}{2\pi}}. \tag{6.11}$$

We want to use these moments to approximate the wavelet coefficients for smooth signals and show that these coefficients will be small when the level j is high. If f is a smooth, twice continuously differentiable signal, then its j, k-wavelet coefficient is

$$b_k^j = \int_{-\infty}^{\infty} f(x)\, 2^{j/2} \psi_2(2^j x - k)\, dx$$

$$= \int_0^{3 \cdot 2^{-j}} f(x + 2^{-j}k)\, 2^{j/2} \psi_2(2^j x)\, dx.$$

If j is large enough, the interval over which we are integrating will be small, and we may then replace $f(x + 2^{-j}k)$ by its quadratic Taylor polynomial in x, $f(x + 2^{-j}k) \approx f(2^{-j}k) + xf'(2^{-j}k) + \frac{1}{2}x^2 f''(2^{-j}k)$. Doing this, we get the following approximation for b_k^j:

$$b_k^j \approx \int_0^{3 \cdot 2^{-j}} \left(f(2^{-j}k) + xf'(2^{-j}k) + \frac{1}{2}x^2 f''(2^{-j}k) \right) 2^{j/2} \psi_2(2^j x)\, dx. \tag{6.12}$$

The integral on the right can be reduced to doing the integrals for the first three moments of ψ_2. Since the first two moments vanish and the third is given in (6.11), we can evaluate the integral. (See Exercise 3.) Our final result is that the j, k-wavelet coefficient is approximately

$$b_k^j \approx -\frac{1}{16}\sqrt{\frac{3}{2\pi}}\, 2^{-5j/2} f''(2^{-j}k). \tag{6.13}$$

Singularity Detection. As an application of the formula in (6.13), we find a point where an otherwise smooth function has a discontinuity in its derivative. This application is called *singularity detection,* and this process can be used, among other things, to detect cracks in material.

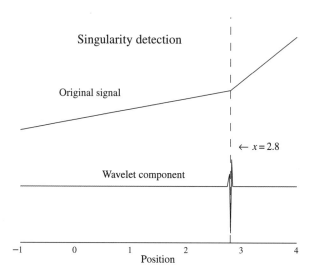

Figure 4 The original signal is piecewise linear on the interval $[-1, 4]$, with a break at $x = 2.8$. The wavelet part comes from a first-order decomposition using the $N = 2$ wavelet; it is essentially 0, except for the interval near $x = 2.8$.

As an example, we will determine where the piecewise linear function shown in Figure 4 changes slope. Keep in mind that the formula (6.13) is exact in any region where a signal f is constant, linear, or quadratic in x, because the quadratic Taylor polynomial for f *is* f in these cases. Since $f'' = 0$ where f is linear, (6.13) implies that the only *non*zero wavelet coefficients will come from a small region near the corner point where the slope changes. The signal itself has the form

$$f(x) := \begin{cases} 0.37x + 1.37, & -1 \le x \le 2.8 \\ 1.58x - 2.03, & 2.8 \le x \le 4. \end{cases}$$

After sampling the signal at 256 equally spaced points, we decomposed it using the $N = 2$ Daubechies wavelet. Thus, our starting level was $j = 8$; the level down from this is $j = 7$. The only appreciable wavelet coefficient was $b_{97}^7 \approx -0.01$. The rest were on the order of 10^{-14}. The index $k = 97$ corresponds to a wavelet supported on the x-interval $[2.79, 2.83]$; consequently, the singularity is located in this interval.

6.3   Computational Issues

Consider the following problem. We want to use $N = 2$ Daubechies wavelets to decompose a signal that we have n samples for, $s_0 \ldots, s_{n-1}$. These samples will be regarded as part of our top-level approximation coefficient sequence, a^j. We will filter these numbers using the high- and low-pass filters for the Daubechies wavelets. These are H and L, and they have impulse responses given by the sequences h and ℓ, where

$$\ell = \frac{1}{2}\left(\cdots 0 \quad \underbrace{p_3 \quad p_2 \quad p_1 \quad p_0}_{k=-3\cdots 0} \quad 0 \quad 0 \quad \cdots\right) \quad \ell_k = \frac{1}{2}\overline{p}_{-k}$$

$$h = \frac{1}{2}\left(\cdots 0 \quad 0 \quad 0 \quad \underbrace{-p_0 \quad p_1 \quad -p_2 \quad p_3}_{k=-1\cdots 2} \quad \cdots\right) \quad h_k = \frac{1}{2}(-1)^k p_{k+1}.$$

We convolve ℓ and h with a^j and downsample. Thus, for any a^j—not just our signal with eight samples—we have

$$a_k^{j-1} = D(\ell * a^j)_k = \frac{1}{2}\left(p_0 a_{2k}^j + p_1 a_{2k+1}^j + p_2 a_{2k+2}^j + p_3 a_{2k+3}^j\right), \qquad (6.14)$$

$$b_{k+1}^{j-1} = D(h * a^j)_k = \frac{1}{2}\left(p_3 a_{2k}^j - p_2 a_{2k+1}^j + p_1 a_{2k+2}^j - p_0 a_{2k+3}^j\right). \qquad (6.15)$$

To compute each level $j-1$ coefficient, we need four consecutive samples, always starting at an even number. For example, if we have $n = 8$ samples, $s_0, \ldots, s_7$, then to compute the a_2^2 coefficients, we need s_4 through s_7. To compute a_3^2, we need s_6 through s_9. But we do not have s_9! It's no surprise that at some point we will run out of coefficients to filter. The surprise is that we are only able to compute $k = 0$, 1, 2. This means that from eight samples, we only get three decomposition coefficients, not the four that we would expect. This is the *overspill* problem. The filters H and L spill over to samples we don't have.

   What to do? First, the overspill problem comes about because we don't know what the signal was before or after our set of samples. Thus we need to extend the signal in some fashion beyond our initial set of samples. Here are a few of the standard choices. In the accompanying illustrations, the filled circles represent the original signal, and the open circles are points in the signal's extension.

Zero padding. Extend the signal by adding zeros at the two ends. This is appropriate if the signal is very long, and the ends of the signal do not matter, or if the signal really is abruptly started and stopped. This amounts to setting $s_k = 0$ if $k < 0$ or $k > n - 1$.

Periodic extension. Another approach is to reuse the sampled data by making the signal periodic, so that $s_{k+n} = s_k$. For example, with eight samples, $s_0, \ldots, s_7$, we would let $s_8 = s_0$, $s_9 = s_1$, and so on.

Smooth padding. Extend the signal by linearly extrapolating the data near the two ends. This is appropriate if signal isn't too noisy, at least near the ends.

Symmetric extensions. The signal is evenly extended at the endpoints by reflection. There are two ways to do this. The signal can be reflected about a line through an endpoint, or about a line midway between an endpoint and the next point on the grid. The first type is illustrated at the left endpoint, and the second at the right one.

6.4   The Scaling Function at Dyadic Points

Although the values of the scaling function ϕ do not enter into the decomposition and reconstruction algorithms, it is useful to be able to compute approximate values of ϕ in order to verify some of its properties (like continuity). An iterative method for computing ϕ is given in Theorem 5.23. However, this algorithm is somewhat cumbersome from a computational point of view. A more computationally efficient method for computing the scaling function ϕ is to use the scaling equation to compute the value of ϕ at all *dyadic* values, $x = l/2^n$, where l and n are integers. We illustrate this process in the following steps. To simplify the exposition, we concentrate on the Daubechies case ($N = 2$) with four nonzero p-coefficients as constructed in Section 6.1, but the procedure easily generalizes.

Step 1. *Compute ϕ at all the integer values.*

Let $\phi_l = \phi(l)$ for $l \in Z$. The Daubechies $(N = 2)$ scaling function is nonzero only on the interval $0 < x < 3$ and so $\phi_0 = \phi(0) = 0 = \phi(3) = \phi_3$ (see Figure 2). In particular, ϕ_1 and ϕ_2 are the only nonzero unknown values of ϕ at integer points. In order to arrange the normalization $\int \phi = 1$, we require $\sum_l \phi_l = 1$ or, in our specific case,

$$\phi_1 + \phi_2 = 1. \tag{6.16}$$

Now we use the scaling equation $\phi(x) = \sum_k p_k \phi(2x - k)$. For $x = 1$, this equation becomes

$$\phi_1 = \sum_{k=0}^{3} p_k \phi(2 - k) = p_0 \phi_2 + p_1 \phi_1 \quad \text{(because } \phi_0 = \phi_3 = 0\text{)}.$$

At $x = 2$, the scaling equation becomes

$$\phi_2 = \sum_{k \in Z} p_k \phi(4 - k) = p_2 \phi_2 + p_3 \phi_1.$$

These two equations can be written in matrix form as

$$\begin{pmatrix} \phi_1 \\ \phi_2 \end{pmatrix} = \begin{pmatrix} p_1 & p_0 \\ p_3 & p_2 \end{pmatrix} \begin{pmatrix} \phi_1 \\ \phi_2 \end{pmatrix}.$$

Here, the p-values are known [see (6.5)] and we are solving for the unknowns ϕ_1 and ϕ_2. In order for this matrix equation to have a nonzero solution, the matrix must have an eigenvalue equal to 1. Then (ϕ_1, ϕ_2) would be the corresponding eigenvector with $\phi_1 + \phi_2 = 1$. To find this eigenvector, we rewrite the matrix equation as

$$\begin{pmatrix} p_1 - 1 & p_0 \\ p_3 & p_2 - 1 \end{pmatrix} \begin{pmatrix} \phi_1 \\ \phi_2 \end{pmatrix} = \begin{pmatrix} 0 \\ 0 \end{pmatrix}.$$

A nonzero solution to this matrix equation exists if the first row is a multiple of the second. From Theorem 5.9, $\sum p_{\text{odd}} = \sum p_{\text{even}} = 1$. Therefore the first row is the negative of the second. The equation corresponding to the first row is

$$(p_1 - 1)\phi_1 + p_0 \phi_2 = 0.$$

We restate the normalization equation (6.16):

$$\phi_1 + \phi_2 = 1.$$

These two equations can be solved simultaneously. In the case of Daubechies, the solution is

$$\phi_1 = \frac{1 + \sqrt{3}}{2} \approx 1.366 \qquad \phi_2 = \frac{1 - \sqrt{3}}{2} \approx -0.366.$$

The ϕ-values at all other integer points are zero. Therefore, all the $\phi_l = \phi(l)$, $l \in Z$ have now been determined.

Step 2. *The values of ϕ at the half-integers.*
To compute $\phi(l/2)$, we use the scaling equation

$$\phi(x) = \sum_{k \in Z} p_k \phi(2x - k) \tag{6.17}$$

at $x = l/2$ to obtain

$$\phi(l/2) = \sum_{k=0}^{3} p_k \phi(l - k). \tag{6.18}$$

Since $\phi(l - k)$ is known from Step 1, $\phi(l/2)$ can be computed. We need only compute $\phi(l/2)$ for $l = 1, 2, 3, 4$, and 5 since $\phi(x) = 0$ for $x \le 0$ and $x \ge 3$. When $l = 2$ or 4, $l/2$ is an integer and the values of ϕ at these points are known from Step 1. Thus we need only compute $\phi(l/2)$ for $l = 1, 3$, and 5. Equation (6.18) for $l = 1, 3, 5$ becomes

$$\phi(1/2) = p_0\phi_1 = \frac{(1 + \sqrt{3})^2}{8} \approx .933 \quad (l = 1)$$

$$\phi(3/2) = p_1\phi_2 + p_2\phi_1 = 0 \quad (l = 3)$$

$$\phi(5/2) = p_3\phi_2 = \frac{(-1 + \sqrt{3})^2}{8} \approx .067 \quad (l = 5).$$

The condition $\sum_l \phi(l) = 1$ (from Step 1) implies $\sum_l \phi(l/2) = 2$. Indeed, equation (6.17) gives

$$\sum_{l \in Z} \phi(l/2) = \sum_{l \in Z} \sum_{k=0}^{3} p_k \phi(l - k) = \sum_{k=0}^{3} p_k \sum_{l \in Z} \phi(l - k).$$

By the change of index, the inner sum is $\sum_l \phi(l - k) = \sum_l \phi(l)$. This sum equals 1 from Step 1. By Theorem 5.9, $\sum_k p_k = 2$. Therefore,

$$\sum_{l \in Z} \phi(l/2) = 2$$

as claimed.

Step 3. *Iterate.*
The values of ϕ at the quarter-integers, $l/4$, can be computed from the values of ϕ at the half-integers by letting $x = l/4$ in the scaling equation $\phi(x) = \sum_k p_k \phi(2x - k)$. In general, once ϕ has been computed at the values $x = l/2^{n-1}$, then we can compute the value of ϕ at the values $x = l/2^n$ by inserting $x = l/2^n$ into the scaling equation (6.17):

$$\phi(l/2^n) = \sum_{k \in Z} p_k \phi(l/2^{n-1} - k) = \sum_{k \in Z} p_k \phi\left(\frac{l - 2^{n-1}k}{2^{n-1}}\right).$$

The right side involves values of ϕ at $x = l'/2^{n-1}$ that have been computed from the previous step.

We claim

$$\sum_{l \in Z} \phi(l/2^n) = 2^n. \qquad (6.19)$$

We have already shown this equation for $n = 0$ (Step 1) and $n = 1$ (Step 2). Suppose by induction that we assume this equation is true for $n - 1$. We now show it to be true for n. We have

$$\sum_{l \in Z} \phi(l/2^n) = \sum_{l \in Z} \sum_{k \in Z} p_k \phi(l/2^{n-1} - k) \quad \text{from (6.17)}$$

$$= \sum_{k \in Z} p_k \sum_{l \in Z} \phi(l/2^{n-1} - k) \quad \text{(switch order of summation)}$$

$$= \sum_{k \in Z} p_k \sum_{l \in Z} \phi\left(\frac{l - 2^{n-1}k}{2^{n-1}}\right)$$

$$= \sum_{k \in Z} p_k \sum_{l' \in Z} \phi(l'/2^{n-1}) \quad \text{(with } l' = l - 2^{n-1}k\text{)}.$$

By the induction hypothesis, the inner sum is 2^{n-1} [equation (6.19), with n replaced by $n - 1$]. Therefore,

$$\sum_{l \in Z} \phi(l/2^n) = \sum_{k \in Z} p_k 2^{n-1}.$$

Since $\sum_k p_k = 2$ (Theorem 5.9), the right side equals 2^n, as desired.

As n gets larger, the set of dyadic points $\{l/2^n, \ l \in Z\}$ gets denser. Since any real number is a limit of dyadic points and since the Daubechies scaling function is continuous, the value of ϕ at any value x can be obtained as a limit of ϕ-values at Dyadic points.

The scaling function, ϕ, that is constructed in this manner satisfies the normalization condition $\int \phi = 1$. To see this, we regard $\int \phi \, dx$ as a limit as $n \to \infty$ of a Riemann sum over a partition given by $\{x_l = l/2^n; \ l = \ldots, -1, 0, 1, 2, \ldots\}$. The width of this partition is $\Delta x = 1/2^n$. So we have

$$\int_{-\infty}^{\infty} \phi(x) \, dx = \lim_{n \to \infty} \sum_{l \in Z} \phi(x_l) \, \Delta x = \lim_{n \to \infty} \sum_{l \in Z} \phi(l/2^n) \, (1/2^n).$$

Since $\sum_l \phi(l/2^n) = 2^n$, the right side is 1, as claimed.

6.5 Exercises

1. Show that $\widetilde{\phi}(x) := \phi_2(\frac{3}{2} - x)$ satisfies the $N = 2$ Daubechies scaling relation, but with the coefficients reversed; that is,

$$\widetilde{\phi}(x) = p_3 \widetilde{\phi}(2x) + p_2 \widetilde{\phi}(2x - 1) + p_1 \widetilde{\phi}(2x - 2) + p_0 \widetilde{\phi}(2x - 3).$$

What is the corresponding equation for $\widetilde{\psi}(x) := \psi_2(\frac{3}{2} - x)$? Sketch both $\widetilde{\phi}$ and $\widetilde{\psi}$.

2. Show that equation (6.9) holds.

3. Use the first three moments of ψ_2 to evaluate integral in (6.12) and thus obtain the approximation to b_k^j given in (6.13).

4. Repeat Exercises 9 and 10 in Chapter 4 using Daubechies ($N = 2$) wavelets.

5. Repeat Exercise 4 for the signal

$$g(t) = -52t^4 + 100t^3 - 49t^2 + 2 + N(100(t - 1/3)) + N(200(t - 2/3))$$

where

$$N(t) = te^{-t^2}.$$

Compare your results with that of Daubechies wavelets with those of Exercise 9 in Chapter 3.

6. Let $f(t)$ be defined via

$$f(t) = \begin{cases} 0 & t < 0, t > 1 \\ t(1 - t) & 0 \leq t \leq 1. \end{cases}$$

Sample f at the dyadic points $k \times 2^{-8}$, $k = -256, \ldots, 512$. Thus, $j = 7$ is the top level. Using the Daubechies ($N = 2$) ψ_2 wavelet, implement a 1-level decomposition. Plot the magnitudes of the level 7 wavelet *coefficients*. Which wavelet has the largest coefficient? What t corresponds to this wavelet? Explain.

7. Let g be defined by

$$g(t) = \begin{cases} 0 & t < 0, t > 1 \\ t^2(1 - t)^2 & 0 \leq t \leq 1. \end{cases}$$

Sample g at the dyadic points $k \times 2^{-8}$, $k = -256, \ldots, 512$ ($j = 7$ is the top level). Using the Daubechies ($N = 2$) ψ_2 wavelets, implement a 1-level decomposition. Plot the magnitudes of the level 7 wavelet *coefficients* for each, and determine which wavelet coefficient is greatest. Repeat with the ψ_3 ($N = 3$) wavelet. Give an explanation of why the greatest coefficients occur where they do.

8. (*Polynomial Suppression.*) Frequently signals have a "bias," which is a polynomial part (usually linear or quadratic) added to a bounded, rapidly oscillating part—$f(t) = 2 + 5t + t^2 + \frac{1}{50} \sin(64\pi t) \cos(6\pi t)$, for example. Suppose that we have taken 1024 samples of f on the interval $[-1, 1]$. Come up with a strategy for separating the two signals via a Daubechies wavelet analysis. Use MATLAB to carry it out. (*Hint*: What is the smallest value of N needed to reproduce quadratics exactly? Also, smooth padding works best for these examples.)

9. (*Noise Removal.*) Wavelets can be used to remove noise from a signal. Let $f(t) = \sin(8\pi t)\cos(3\pi t) + n(t)$, with $n(t)$ being the noise. Numerically, we can model $n(t)$ by using a random number generator, such as MATLAB's **rand**. Take 1500 samples on $[-2, 3]$ of f and do an analysis with $N = 2$, 3, and 6 Daubechies wavelets. Experiment with different levels. Which wavelet does the best job?

Chapter 7

Other Wavelet Topics

This chapter contains a variety of topics of a more advanced nature that relate to wavelets. Since these topics are more advanced, an overview is presented with details left to the various references.

7.1 Computational Complexity

7.1.1 Wavelet Algorithm

In this section, we briefly discuss the number of operations required to compute the wavelet decomposition of a signal (the computational complexity of the wavelet decomposition algorithm).

To take a concrete setting, suppose f is a continuous signal defined on the unit interval $0 \le t \le 1$. Let

$$a_l^n = f(l/2^n) \quad l = 0, \ldots, 2^n - 1$$

be a discrete sampling of f. We wish to count the number of multiplications number $N = 2^n$.

The decomposition algorithm stated in equation (5.12) or equation (5.17) computes the a^{j-1} and b^{j-1}, $j = n, \ldots, 1$ using the formulas

$$a_l^{j-1} = (1/2) \sum_k \overline{p}_{k-2l} a_k^j$$

$$b_l^{j-1} = (1/2) \sum_k (-1)^k p_{1-k+2l} a_k^j.$$

Note that there are half as many a^{j-1} coefficients as there are a^j coefficients. So the index l on the left runs from 0 to $2^{j-1} - 1$. Let L be the number of nonzero values of p_k. For Haar, $L = 2$, and for the Daubechies system constructed in the previous chapter, $L = 4$. Thus, there are 2^{j-1} coefficients at level $j - 1$ (for a_l^{j-1}) to be computed, each of which requires $L + 1$ multiplications (including the multiplication by $1/2$). The same number of computations are required for b_l^{j-1}. As j runs from n to 1, the total number of multiplications required for decomposition is

$$2(L+1)2^{n-1} + \cdots + 2(L+1)2^0 = 2(L+1) \sum_{j=0}^{n-1} 2^j$$
$$= 2(L+1)(2^n - 1) \quad \text{using equation (1.23)}$$
$$\approx 2(L+1)N.$$

This result is often summarized by stating that the wavelet decomposition algorithm requires $O(N)$ multiplication operations, where N is the number of data at the top level. Here, $O(N)$ stands for a number that is proportional to N. The proportionality constant in this case is $2(L + 1)$ (which is 6 in the case of Haar or 10 in the case of Daubechies).

By comparison, the number of multiplicative operations for the fast Fourier transform is $O(N \log N)$ (see Section 3.1.3). However, this comparison is unfair since the fast Fourier transform decomposes a signal into all its frequency components between 0 and $N/2$. By contrast, the wavelet decomposition algorithm decomposes a signal into its frequency components that lie in ranges bounded by powers of 2. For example,

$$w_{j-1}(x) = \sum b_l^{j-1} \psi(2^{j-1}x - l)$$

is the decomposition of f given in Theorem 5.11. This should be regarded as the component of f that involves frequencies in the range 2^{j-1} to 2^j. Therefore, the wavelet decomposition algorithm does not decompose the signal into the more detailed range of frequencies that are offered by the discrete Fourier transform.

7.1.2 Wavelet Packets

Wavelets can be used to obtain a decomposition of a signal into a finer gradation of frequency components using *wavelet packets*. We briefly describe the technique for doing this, which requires $O(N \log N)$ operations; therefore, it is no more efficient than the fast Fourier transform. A more complete discussion may be found in [13, Chapter 7].

We illustrate the idea of wavelet packets in the specific when $n = 3$. A signal $f_3 \in V_3$ can be decomposed via Theorem 5.11 as

$$f_3 = w_2 + w_1 + w_0 + f_0 \in W_2 \oplus W_1 \oplus W_0 \oplus V_0.$$

We can put this decomposition into a more hierarchical structure.

$$f_3 \xrightarrow{\ L\ } f_2 \xrightarrow{\ L\ } f_1 \xrightarrow{\ L\ } f_0$$

with H branches leading to w_2, w_1, w_0.

The arrows marked H are the projection operators onto the wavelet components. The arrows marked L stand for the projection to the next lower level (e.g., the projection from V_2 to V_1) The H stands for "high-pass filter" since the wavelet part of a particular level represents the component with higher frequency (e.g., the W_1 component of $V_2 = W_1 \oplus V_1$ represents higher-frequency components than the V_1 component). Likewise, the L stands for "low-pass filter" since it projects to the next lower level.

The f_3 contains frequency components from 1 to 8. The w_2 component contains frequency components from 5 to 8. The analogous frequency ranges for the w_1, w_0 and f_0 components are 3 to 4, 2, and 1, respectively. The idea behind wavelet packets is to apply the high-pass and low-pass filters to each of w_2, w_1, and w_0. The high-pass filter and low-pass filters applied to w_2 create new components w_{21} and w_{20}, respectively, which are orthogonal to one another. The process is iterated until a complete decomposition into all frequencies is obtained.

In the case of f_3, a complete decomposition into all components is illustrated in the following diagram.

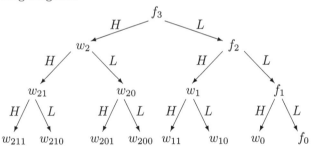

To count the number of operations required, imagine this diagram were extended to handle n levels ($n = 3$ is illustrated). Each component in the kth level of the diagram requires $O(2^{N-k})$ computations. Since there are 2^k components in the kth level, the total number of computations for this level is $O(2^n)$ (independent of k). Since there are n levels (not counting the first, which only contains the original signal), the total number of operations is $O(n2^n)$ or $N \log_2 N$, where $N = 2^n$. This is the same number of operations required by the fast Fourier transform.

7.2 Wavelets in Higher Dimensions

Signal processing with wavelets of video images requires a generalization of wavelets to several dimensions (two are required for a static pictorial image). Wavelets in higher dimensions are obtained by taking the *tensor product* of

one-dimensional wavelets. We briefly explain this concept in the case of two
dimensions, which we think of as R^2 with coordinates (x, y).

Suppose ϕ and ψ are the scaling function and wavelet for a multiresolution
analysis (for example, the Haar system or the one constructed by Daubechies).
As shown in previous sections, the functions

$$\phi_{jl}(x) = 2^{j/2}\phi(2^j x - l) \qquad \psi_{jl}(x) = 2^{j/2}\psi(2^j x - l)$$

are orthonormal bases for the spaces V_j and W_j, respectively. Moreover, each
ψ_{jl} is orthogonal to all the ϕ_{jl}. For each set of indices j, l, j', l', define the
functions

$$\phi_{j,l,j',l'}(x, y) = \phi_{jl}(x)\phi_{j'l'}(y) \quad \text{and} \quad \psi_{j,l,j',l'}(x, y) = \psi_{jl}(x)\psi_{j'l'}(y). \quad (7.1)$$

The indices j and j' vary between 0 and n (the top level) as before. The
indices l and l' correspond to the translational components and depend on the
domain of interest. For example, if the signal is defined on the unit square
$\{(x, y);\ 0 \le x, y \le 1\}$, then $0 \le l \le 2^j - 1$ and $0 \le l' \le 2^{j'} - 1$. As an easy
exercise, you can show that for each pair of indices j, j', the sets of functions
$\{\phi_{j,l,j',l'}(x, y);\ l, l' \in Z\}$ and $\{\psi_{j,l,j',l'}(x, y);\ l, l' \in Z\}$ are orthonormal bases
for $V_j \otimes V_{j'}$ and $W_j \otimes W_{j'} \subset L^2(R^2)$. Here, $V_j \otimes V_{j'}$ is called the *tensor product*
of V_j and $V_{j'}$ and is the space generated by $f(x)g(y)$ where $f \in V_j$ and $g \in V_{j'}$.

A signal $f \in L^2(R^2)$ can be discretized in both dimensions. For example, if
the domain of the signal is the unit square $\{(x, y) \in R^2; 0 \le x, y \le 1\}$, then we
discretize by setting

$$f_{n,n}(x, y) = \sum_{k,k'} a_{k,k'}^{n,n}\phi(2^n x - k)\phi(2^n y - k') \quad \text{where } a_{k,k'}^{n,n} = f(k/2^n, k'/2^n).$$

The decomposition and reconstruction algorithms are pieced together in each
variable separately. Although a precise algorithm is somewhat tedious, the
ideas can be simply explained. Equation (5.10) can be used to decompose the
functions $\phi(2^n x - k)$ and $\phi(2^n y - k')$ into a linear combination of $\phi(2^{n-1}x - l)$,
$\psi(2^{n-1}x - l)$, $\phi(2^{n-1}y - l)$ and $\psi(2^{n-1}x - l)$. The signal $f_{n,n}$ becomes

$$f_{n,n} = w_{n-1,n-1} + fab_{n-1,n-1} + fba_{n-1,n-1} + faa_{n-1,n-1}$$

where

$$w_{n-1,n-1} \text{ is a linear combination of } \psi(2^{n-1}x - l)\psi(2^{n-1}y - l')$$

$$fab_{n-1,n-1} \text{ is a linear combination of } \phi(2^{n-1}x - l)\psi(2^{n-1}y - l')$$

$$fba_{n-1,n-1} \text{ is a linear combination of } \psi(2^{n-1}x - l)\phi(2^{n-1}y - l')$$

$$faa_{n-1,n-1} \text{ is a linear combination of } \phi(2^{n-1}x - l)\phi(2^{n-1}y - l').$$

The terms, fab, fba, and faa are further decomposed until only $\psi(x - l)$ or
$\phi^0(x - l)$ terms appear. After collecting terms,

$$f_{n,n} = \sum_{j,k=-1}^{n-1} w_{j,k}$$

where $w_{j,k} \in W_j \otimes W_k$ for $0 \le j, k \le n-1$, $w_{j,-1} \in W_j \otimes V_0$ and $w_{-1,j} \in V_0 \otimes W_j$.

Once decomposed, a signal can be filtered or compressed just as in the one-dimensional case. Reconstruction is accomplished by combining the reconstruction algorithms in both variables. The following equations

$$\phi(x) = \sum_k p_k \phi(2x - k)$$

$$\psi(x) = \sum_k (-1)^k \overline{p}_{1-k} \phi(2x - k)$$

along with the analogous equations in the y-variable can be used to express $w_{-1,-1}$ and $w_{0,0}$ in terms of $\phi(2x - k)\phi(2y - k') \in V_1 \otimes V_1$. Replacing x (and y) by $2^j x$ (and $2^{j'} y$) in the preceding equations allows any element in $W_j \otimes W_{j'}$ to be expressed in terms of elements in $V_{j+1} \otimes V_{j'+1}$. This process is continued until the signal is in the form

$$f_{n,n}(x,y) = \sum_{k,k'} a^{n,n}_{k,k'} \phi(2^n x - k)\phi(2^n y - k')$$

where, of course, the $a^{n,n}_{k,k'}$ have now changed due to the filtering or compression process. The value, $a^{n,n}_{k,k'}$, represents the value of the filtered or compressed signal at $x = k/2^n$, $y = k'/2^n$, $0 \le k, k' \le 2^n - 1$.

7.3 Relating Decomposition and Reconstruction

In this section we analyze the decomposition and reconstruction algorithms and we show that the latter involves the l^2-adjoint of the former. As before, we will assume that ϕ and ψ are the scaling and wavelet functions associated to a multiresolution analysis. We rewrite the decomposition and reconstruction algorithms using the orthonormal bases $\phi_{jk}(x) = 2^{j/2}\phi(2^j x - k) \in V_j$ and $\psi_{jk}(x) = 2^{j/2}\psi(2^j x - k) \in W_j$.

The decomposition algorithm (5.9) becomes the following:

THEOREM 7.1 (Decomposition) *A function $f = \sum_l a^n_l \phi_{nl}(x)$ can be decomposed as*

$$f = w_{n-1} + w_{n-2} + \cdots + w_0 + f_0$$

with

$$w_{j-1}(x) = \sum_l b^{j-1}_l \psi_{j-1,l}(x)$$

where

$$a^{j-1}_l = \frac{1}{\sqrt{2}} \sum_k \overline{p_{k-2l}}\, a^j_k$$

$$b^{j-1}_l = \frac{1}{\sqrt{2}} \sum_k (-1)^k p_{1-k+2l}\, a^j_k.$$

In terms of convolution and the downsampling operator D,

$$a^{j-1} = D(P * a^j) \quad \text{where } P_k = \frac{1}{\sqrt{2}} \bar{p}_{-k}$$

$$b^{j-1} = D(Q * a^j) \quad \text{where } Q_k = \frac{1}{\sqrt{2}}(-1)^k p_{1+k}.$$

The reconstruction algorithm (5.11) can be restated as follows.

THEOREM 7.2 (Reconstruction) *Suppose*

$$f_{j-1}(x) = \sum_l a_l^{j-1} \phi_{j-1,l}(x) \qquad w_{j-1}(x) = \sum_l b_l^{j-1} \psi_{j-1,l}(x)$$

and let

$$f_j(x) = f_{j-1}(x) + w_{j-1}(x) = \sum_l a_l^j \phi_{jl}(x).$$

Then the coefficients a_l^j can be computed from the sequences a^{j-1} and b^{j-1} by

$$a_l^j = \frac{1}{\sqrt{2}} \sum_k a_k^{j-1} p_{l-2k} + \frac{1}{\sqrt{2}} \sum_k b_k^{j-1} \bar{p}_{1+2k-l}(-1)^l.$$

In terms of convolution and the upsampling operation U,

$$a^j = P^* * (U a^{j-1}) + Q^* * (U b^{j-1})$$

where $P_k^* = \frac{1}{\sqrt{2}} p_k$ and $Q_k^* = \frac{1}{\sqrt{2}} \bar{p}_{1-k}(-1)^k$.

Except for the factor of $\frac{1}{\sqrt{2}}$, which results from using the orthonormal bases ϕ_{jl} and ψ_{jl}, these theorems are identical to the algorithms in (5.12) and (5.13) [or (5.17) and (5.22)].

Recall that l^2 is the space of all sequences $x = (x_j)$; $j \in Z$ with $\sum_j |x_j|^2 < \infty$. In practical applications, there are only a finite number of nonzero x_j and so the condition $\sum_j |x_j|^2 < \infty$ is automatically satisfied. The key operators involved with decomposition algorithm are the downsampling operator $D: l^2 \mapsto l^2$ and the operators $T_P: l^2 \mapsto l^2$ and $T_Q: l^2 \mapsto l^2$:

$$T_P(x) = P * x \quad x \in l^2$$
$$T_Q(x) = Q * x \quad x \in l^2$$

where P and Q are the sequences defined in Theorem 7.1. Likewise, the key operators in the reconstruction algorithm are the upsampling operator, U, and the operators $T_{P^*}: l^2 \mapsto l^2$ and $T_{Q^*}: l^2 \mapsto l^2$:

$$T_{P^*}(x) = P^* * x \quad x \in l^2$$
$$T_{Q^*}(x) = Q^* * x \quad x \in l^2$$

where P^* and Q^* are the sequences defined in Theorem 7.2.

Our goal now is to show that the operators T_P and T_{P^*} are adjoints of one another and likewise for the operators involving Q. Recall that if $T \colon V \mapsto V$ is a linear operator on an inner product space, V, then its adjoint T^* is defined by

$$\langle Tv, w \rangle = \langle v, T^*w \rangle.$$

LEMMA 7.3

- *The adjoint of T_P is T_{P^*}.*

- *The adjoint of T_Q is T_{Q^*}.*

- *The adjoint of D (the downsampling operator) is U (the upsampling operator).*

Proof The first two properties follow from Theorem 3.14, which states that the adjoint of the operator associated with the convolution with a sequence f_n is the convolution operator associated with the sequence $f_n^* = \overline{f}_{-n}$. Thus, the first property follows from the fact that $P_k^* = \frac{1}{\sqrt{2}}p_k = \overline{P}_{-k}$. Likewise, $Q_k^* = \frac{1}{\sqrt{2}}\overline{p}_{1-k}(-1)^k = \overline{Q}_{-k}$ and so the second property follows from Theorem 3.14.

For the third property, let x and y be sequences in l^2. Using the definition of D,

$$\langle Dx, y \rangle_{l^2} = \sum_n (Dx)_n \overline{y}_n = \sum_n x_{2n} \overline{y}_n.$$

On the other hand, only the even entries of Uy are nonzero and $(Uy)_{2n} = y_n$ (by definition). Therefore,

$$\langle x, Uy \rangle = \sum_n x_{2n} \overline{(Uy)}_{2n} = \sum_n x_{2n} \overline{y}_n.$$

By comparing these two sets of equations, we conclude $\langle Dx, y \rangle = \langle x, Uy \rangle$ as desired. ◆

Using the lemma and Theorems 7.1 and 7.2, the decomposition and reconstruction algorithms can be restated (essentially) as being adjoints of each other.

THEOREM 7.4 (Decomposition) *Let $P_k = \frac{1}{\sqrt{2}}\overline{p}_{-k}$ and $Q_k = \frac{1}{\sqrt{2}}(-1)^k p_{1+k}$. Suppose*

$$T_0 = D \circ T_P \quad and \quad T_1 = D \circ T_Q$$

where D is the downsampling operator and T_P (respectively T_Q) is the operator that convolves with the sequence P (respectively Q). A function $f_j = \sum_l a_l^j \phi_{jl}(x)$ can be decomposed as

$$f = w_{j-1} + f_{j-1}$$

with

$$w_{j-1}(x) = \sum_l b_l^{j-1} \psi_{j-1,l}(x) \quad and \quad f_{j-1} = \sum_l a_l^{j-1} \phi_{j-1,l}(x)$$

where

$$a^{j-1} = T_0(a^j) \quad and \quad b^{j-1} = T_1(a^j). \tag{7.2}$$

Conversely, the sequence a^j can be reconstructed from a^{j-1} and b^{j-1} by

$$a^j = T_0^*(a^{j-1}) + T_1^*(b^{j-1}). \tag{7.3}$$

Proof   The decomposition formulas follow from Theorem 7.1. For the reconstruction formula, Theorem 7.2 implies

$$
\begin{aligned}
a^j &= (T_{P^*} \circ U)(a^{j-1}) + (T_{Q^*} \circ U)(b^{j-1}) \\
&= (T_P^* \circ D^*)(a^{j-1}) + (T_Q^* \circ D^*)(b^{j-1}) \quad \text{(using Lemma 7.3)} \\
&= (D \circ T_P)^*(a^{j-1}) + (D \circ T_Q)^*(b^{j-1}) \quad \text{(using Theorem 0.32)} \\
&= T_0^*(a^{j-1}) + T_1^*(b^{j-1}).
\end{aligned}
$$

Equations (7.2) and (7.3) can be combined as

$$a^j = (T_0^* \circ T_0 + T_1^* \circ T_1)(a^j),$$

which is another way of stating that the reconstruction process is the adjoint of decomposition and that the reconstruction process inverts decomposition [because $(T_0^* \circ T_0 + T_1^* \circ T_1)$ sends the sequence a^j to itself]. Operators with this property have a special name.

DEFINITION 7.5   *A pair of filters, T_P and T_Q (i.e., convolution operators with the sequences P and Q, respectively), is called a* quadrature mirror filter *if the associated maps $T_0 = D \circ T_P$ and $T_1 = D \circ T_Q$ satisfy*

$$(T_0^* \circ T_0)(x) + (T_1^* \circ T_1)(x) = x \tag{7.4}$$

for all sequences $x = (\ldots, x_{-1}, x_0, x_1, \ldots) \in l^2(Z)$.

An equivalent formulation of equation (7.4) is

$$\|T_0 x\|^2 + \|T_1 x\|^2 = \|x\|^2.$$

For if equation (7.4) holds, then

$$
\begin{aligned}
\|x\|^2 &= \langle x, x \rangle \quad \text{(by definition of } \|\cdot\|) \\
&= \langle (T_0^* \circ T_0 + T_1^* \circ T_1)(x), x \rangle \quad \text{(by equation (7.4))} \\
&= \langle T_0 x, T_0 x \rangle + \langle T_1 x, T_1 x \rangle \quad \text{by definition of adjoint} \\
&= \|T_0 x\|^2 + \|T_1 x\|^2.
\end{aligned}
$$

The l^2-norm of a signal $x = (\ldots, x_{-1}, x_0, x_1, \ldots)$, (i.e., $\sum_n x_n^2$) is proportional to the energy of the signal. So a physical interpretation of equation (7.4) is that a quadrature mirror filter preserves the energy of a signal.

7.3.1 Transfer Function Interpretation

Our goal in this section is to transcribe the defining property of a quadrature mirror filter

$$(T_0^* \circ T_0)(x) + (T_1^* \circ T_1)(x) = x \tag{7.5}$$

into a condition on the transfer functions of the filters T_P and T_Q.

First, we recall the definition of a transfer function given in Chapter 3 (just after Theorem 3.13). A sequence $x = (\ldots, x_{-1}, x_0, x_1, \ldots) \in l^2(Z)$ has a discrete Fourier transform $\widehat{x}$ that is a function on the interval $-\pi \le \theta \le \pi$ defined by

$$\widehat{x}(\theta) = \sum_{n=-\infty}^{\infty} x_n e^{-in\theta}.$$

If T_P is a convolution operator with the sequence P_n—that is, $(T_P x)_n = \sum_k P_{n-k} x_k$—then the transfer function of T_P is the function $\widehat{T_P}$ on the interval $-\pi \le \theta \le \pi$ defined by

$$\widehat{T_P}(\theta) = \sum_n P_n e^{-in\theta}.$$

As shown in Theorem 3.13,

$$\widehat{T_P(x)}(\theta) = \widehat{T_P}(\theta)\widehat{x}(\theta). \tag{7.6}$$

Another key property that we will use is the computation of the transfer function of the transpose of a convolution operator; that is,

$$\widehat{F^*}(\theta) = \overline{\widehat{F}(\theta)} \tag{7.7}$$

which is established in Theorem 3.14.

To transcribe equation (7.5) into one involving transfer functions, we take the discrete Fourier transform (the hat of) both sides of this equation. In what follows, we use the notation $\mathcal{F}(x)(\theta) = \widehat{x}(\theta)$ for expressions, x, that are long. We start with the $(T_0^* \circ T_0)(x)$ term on the left side,

$$\mathcal{F}((T_0^* \circ T_0)(x))(\theta) = \mathcal{F}(T_P^* \circ D^* \circ D \circ T_P)(x)(\theta) \tag{7.8}$$

$$= \widehat{T_P^*}(\theta) \cdot \mathcal{F}((D^* \circ D \circ T_P)(x))(\theta) \tag{7.9}$$

$$= \overline{\widehat{T_P}(\theta)} \cdot \mathcal{F}((D^* \circ D \circ T_P)(x))(\theta) \tag{7.10}$$

where the last two equalities use (7.6) and (7.7), respectively. Now if y is the sequence y_n, $n \in Z$, then from the last section, $(Dy)_n = y_n$ for $n = 2k$ an even integer. So

$$(D^*Dy)_n = \begin{cases} y_n & \text{for } n \text{ even, i.e., } n = 2k \\ 0 & \text{for } n \text{ odd.} \end{cases}$$

Therefore,

$$\widehat{D^*Dy}(\theta) = \sum_k y_{2k} e^{-i2k\theta}$$

which can be redescribed as

$$\widehat{D^*Dy}(\theta) = \sum_n y_n (e^{-in\theta} + e^{-in(\theta+\pi)})/2$$

(the terms when n is odd cancel since then $e^{-in\pi} = -1$). Therefore,

$$\widehat{D^*Dy}(\theta) = (\widehat{y}(\theta) + \widehat{y}(\theta + \pi))/2.$$

Applying this equation to the sequence $y_n = F_0(x)_n$ yields

$$\mathcal{F}((D^* \circ D \circ T_P)(x))(\theta) = \left(\widehat{T_P(x)}(\theta) + \widehat{T_P(x)}(\theta + \pi)\right)/2$$

$$= \left(\widehat{T_P}(\theta)\widehat{x}(\theta) + \widehat{T_P}(\theta + \pi)\widehat{x}(\theta + \pi)\right)/2.$$

Inserting this into the end of (7.10), we obtain

$$\mathcal{F}((T_0^* \circ T_0)(x))(\theta) = \overline{\widehat{T_P}(\theta)} \left(\widehat{T_P}(\theta)\widehat{x}(\theta) + \widehat{T_P}(\theta + \pi)\widehat{x}(\theta + \pi)\right)/2.$$

Similarly,

$$\mathcal{F}((T_1^* \circ T_1)(x))(\theta) = \overline{\widehat{T_Q}(\theta)} \left(\widehat{T_Q}(\theta)\widehat{x}(\theta) + \widehat{T_Q}(\theta + \pi)\widehat{x}(\theta + \pi)\right)/2.$$

Adding these two equations and setting the result equal to $\widehat{x}(\theta)$ [as required by equation (7.5)] yields

$$\left(\frac{|\widehat{T_P}(\theta)|^2 + |\widehat{T_Q}(\theta)|^2}{2}\right)\widehat{x}(\theta)$$

$$+ \left(\frac{\overline{\widehat{T_P}(\theta)}\widehat{T_P}(\theta + \pi) + \overline{\widehat{T_Q}(\theta)}\widehat{T_Q}(\theta + \pi)}{2}\right)\widehat{x}(\theta + \pi) = \widehat{x}(\theta).$$

Let $G(\theta)$ be the term inside the first parenthesis on the left and let $H(\theta)$ be the term on the inside of the second parenthesis. The preceding equation now reads

$$G(\theta)\widehat{x}(\theta) + H(\theta)\widehat{x}(\theta + \pi) = \widehat{x}(\theta),$$

which must hold for all sequences x. We first apply this equation when $\widehat{x}(\theta) = 1$, which occurs when x is the sequence given by $x_0 = 1$ and $x_n = 0$ for nonzero n. We obtain

$$G(\theta) + H(\theta) = 1.$$

Now apply the equation when $\widehat{x}(\theta) = e^{i\theta}$, which occurs when x is the sequence given by $x_1 = 1$ and $x_n = 0$ for all other n. After dividing by $e^{i\theta}$, we obtain

$$G(\theta) - H(\theta) = 1.$$

Adding these two equations, we obtain $G(\theta) = 1$. Subtracting these two equations yields $H(\theta) = 0$. Inserting the expressions for G and H, we obtain the following theorem

THEOREM 7.6 *The filters T_P and T_Q form a quadrature mirror filter if and only if*

$$|\widehat{T_P}(\theta)|^2 + |\widehat{T_Q}(\theta)|^2 = 2$$

$$\overline{\widehat{T_P}(\theta)}\widehat{T_P}(\theta + \pi) + \overline{\widehat{T_Q}(\theta)}\widehat{T_Q}(\theta + \pi) = 0$$

or, equivalently, the matrix

$$\begin{pmatrix} \widehat{T_P}(\theta)/\sqrt{2} & \widehat{T_Q}(\theta)/\sqrt{2} \\ \widehat{T_P}(\theta + \pi)/\sqrt{2} & \widehat{T_Q}(\theta + \pi)/\sqrt{2} \end{pmatrix}$$

is unitary.

When the quadrature mirror filters come from a scaling function and wavelet, then the (transpose of the) unitary matrix in the preceding theorem becomes

$$\begin{pmatrix} \overline{p(\theta)} & \overline{p(\theta + \pi)} \\ \overline{q(\theta)} & \overline{q(\theta + \pi)} \end{pmatrix}$$

with $p(\theta) = P(e^{-i\theta})$ and $q(\theta) = Q(e^{-i\theta})$ as defined in Section 5.3.3. This is the same matrix as that given after Theorem 5.21, and we saw that this matrix was unitary from the techniques given in that section.

7.4 Wavelet Transform

In Chapter 2 on the Fourier transform, we discussed the Fourier inversion formula. In this section, we develop an analogous *wavelet transform* and its associated inversion formula. Let us first review the Fourier transform. For a function, f, in $L^2(R)$, its Fourier transform is given by

$$\mathcal{F}(f)(\lambda) = \frac{1}{\sqrt{2\pi}} \int_{-\infty}^{\infty} f(t)e^{-i\lambda t}\, dt.$$

Another notation for the Fourier transform is $\widehat{f}(\lambda)$. The inverse Fourier transform of a function $g \in L^2(R)$ is given by

$$\mathcal{F}^{-1}(g)(x) = \frac{1}{\sqrt{2\pi}} \int_{-\infty}^{\infty} g(\lambda)e^{i\lambda x}\, d\lambda.$$

The content of the Fourier inversion Theorem 2.1 is that $\mathcal{F}^{-1}$, as defined previously, really is the inverse operator of $\mathcal{F}$; that is,

$$f = \mathcal{F}^{-1}\mathcal{F}(f).$$

As discussed in Chapter 2, $\mathcal{F}(f)(\lambda)$ roughly measures the component of f that oscillates with frequency λ. The inversion formula $f = \mathcal{F}^{-1}\mathcal{F}(f)$, when written as

$$f(x) = \frac{1}{\sqrt{2\pi}} \int_{-\infty}^{\infty} \widehat{f}(\lambda) e^{i\lambda x} \, d\lambda,$$

expresses the fact that f can be written as a weighted sum (or integral) of its various frequency components.

7.4.1   Definition of the Wavelet Transform

The wavelet transform and its associated inversion formula also decompose a function into a weighted sum of its various frequency components. However, this time the weights involve a given wavelet rather than the exponential term $e^{i\lambda x}$.

To introduce the wavelet transform, we assume that a wavelet function, $\psi(x)$ is given that satisfies the following two requirements:

1. $\psi(x)$ is continuous and has exponential decay [i.e., $\psi(x) \leq Me^{-C|x|}$ for some constants C and M].

2. The integral of ψ is zero [i.e., $\int_{-\infty}^{\infty} \psi(x) \, dx = 0$].

An example of a suitable wavelet function is $\psi(x) = xe^{-x^2}$, whose graph appears in Figure 1. Another example is the wavelet function of Daubechies constructed in the previous chapter. The Daubechies wavelet is only nonzero on a finite interval, so the first requirement is automatically satisfied (by taking the constant M large enough). Note that no assumptions are made concerning orthogonality of the wavelet with its translates. The Daubechies wavelets are constructed so that their translates are orthogonal, but the preceding example $\psi(x) = xe^{-x^2}$ is not orthogonal to its translates.

In the derivations that follow, we assume that $\psi(x)$ equals zero outside some fixed interval $-A \leq x \leq A$, which is a stronger condition than the first condition just given. However, every derivation given can be modified to include wavelets with exponential decay.

We are now ready to state the definition of the wavelet transform.

DEFINITION 7.7   *Given a wavelet ψ satisfying the two requirements just given, the wavelet transform of a function $f \in L^2(R)$ is a function $W_f : R^2 \mapsto R$ given by*

$$W_f(a, b) = \frac{1}{\sqrt{|a|}} \int_{-\infty}^{\infty} f(x)\overline{\psi\left(\frac{x-b}{a}\right)} \, dx.$$

From the preceding definition, it is not clear how to define the wavelet transform at $a = 0$. However, the change of variables $y = (x - b)/a$ converts the wavelet transform into the following:

$$W_f(a, b) = \sqrt{|a|} \int_{-\infty}^{\infty} f(ya + b))\overline{\psi(y)} \, dy.$$

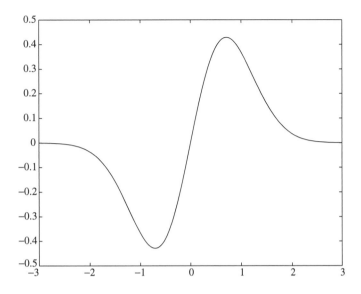

Figure 1 Graph of $\psi_{1,0}(x) = \psi(x) = xe^{-x^2}$

From this representation, clearly $W_f(a, b) = 0$ when $a = 0$.

As a gets small, the graph of

$$\psi_{a,b}(x) = \frac{1}{\sqrt{|a|}} \psi \left(\frac{x - b}{a} \right)$$

becomes tall and skinny, as illustrated in the graphs of $\psi_{1,0}$ and $\psi_{1/2,0}$ with $\psi(x) = xe^{-x^2}$ given in Figures 1 and 2, respectively. Therefore, the frequency of $\psi_{a,b}$ increases as a gets small. Also note that if most of the support of ψ (i.e., the nonzero part of the graph of ψ) is located near the origin (as with the preceding example), then most of the support of $\psi_{a,b}$ will be located near $x = b$. So $W_f(a, b)$ measures the frequency component of f that vibrates with frequency proportional to $1/a$ near the point $x = b$.

7.4.2   Inversion Formula for the Wavelet Transform

The inversion formula for the wavelet transform is given in the following theorem.

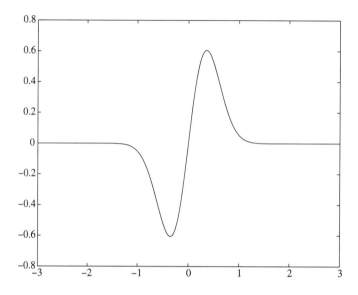

Figure 2 Graph of $\psi_{1/2,0}$

THEOREM 7.8   *Suppose ψ is a continuous wavelet satisfying the following:*

- *ψ has exponential decay at infinity.*
- *$\int_{-\infty}^{\infty} \psi(x)\,dx = 0$.*

Then for any function $f \in L^2(R)$, the following inversion formula holds:

$$f(x) = \frac{1}{C_\psi} \int_{-\infty}^{\infty} \int_{-\infty}^{\infty} |a|^{-1/2} \psi\left(\frac{x-b}{a}\right) W_f(a,b)\, \frac{db\,da}{a^2}$$

where

$$C_\psi = 2\pi \int_{-\infty}^{\infty} \frac{|\widehat{\psi}(\lambda)|^2}{|\lambda|}\, d\lambda.$$

As with the Fourier inversion theorem, the preceding wavelet inversion theorem essentially states that a function f can be decomposed as a weighted sum (or integral) of its frequency components, as measured by $W_f(a,b)$. The difference is that with the wavelet inversion theorem, two parameters, a and b, are involved since the wavelet transform involves a measure of the frequency (using the parameter a) of f near the point $x = b$.

Before we start the proof, we should state why C_ψ is finite (since the integrand defining it appears to become infinite at $\lambda = 0$). First, we split the integral that defines C_ψ into two pieces: one involving an integral over $|\lambda| \geq 1$ and a second involving the integral over $|\lambda| < 1$. For the first integral, we note that ψ has exponential decay and so ψ is a member of $L^2(R)$. Therefore, its

Fourier transform also belongs to $L^2(R)$ and so

$$\int_{|\lambda| \geq 1} \frac{|\widehat{\psi}(\lambda)|^2}{|\lambda|} \, d\lambda \leq \int_{|\lambda| \geq 1} |\widehat{\psi}(\lambda)|^2 \, d\lambda$$

$$< \infty \quad \text{since } \widehat{\psi} \in L^2(R).$$

For the second integral, it is a fact that the exponential decay assumption on ψ implies $\widehat{\psi}$ is differentiable. From a first-order (Taylor) expansion about $\lambda = 0$,

$$\widehat{\psi}(\lambda) = \widehat{\psi}(0) + O(\lambda) \quad \text{(Taylor's theorem)}$$

where $O(\lambda)$ stands for terms that are bounded by $C|\lambda|$ for some constant C. We also have $\widehat{\psi}(0) = \int \psi(x) \, dx = 0$ by the second assumption on ψ stated in the theorem. Therefore, the second integral can be estimated by

$$\int_{|\lambda| < 1} \frac{|\widehat{\psi}(\lambda)|^2}{|\lambda|} \, d\lambda \leq \int_{|\lambda| < 1} \frac{|C\lambda|^2}{|\lambda|} \, d\lambda < \infty.$$

Since both integral pieces of C_ψ are finite, we conclude that C_ψ is finite as well.

Proof of the Wavelet Transform Theorem   Let $F(x)$ be the quantity given on the right of the main statement of the theorem; that is,

$$F(x) = \frac{1}{C_\psi} \int_{-\infty}^{\infty} \int_{-\infty}^{\infty} |a|^{-1/2} \psi\left(\frac{x-b}{a}\right) W_f(a,b) \frac{db \, da}{a^2}.$$

We must show that $F(x)$ is equal to $f(x)$. We accomplish this using the following two steps.

- **Step 1.** We first show

$$F(x) = \frac{1}{C_\psi} \int_{-\infty}^{\infty} \frac{da}{\sqrt{|a|}a^2} \int_{-\infty}^{\infty} \overline{\mathcal{F}_b\left\{\psi\left(\frac{x-b}{a}\right)\right\}}(y) \mathcal{F}_b\{W_f(a,b)\}(y) \, dy$$

where $\mathcal{F}_b\{\cdot\}$ stands for the Fourier transform of the quantity inside the brackets $\{,\}$ with respect to the variable b.

   This step follows by applying Plancherel's formula (Theorem 2.12), which states that $\int vu = \int \mathcal{F}(\overline{v})\mathcal{F}(u)$. This theorem is applied to the b-integral occurring in the definition of $F(x)$ and where $v(b) = \psi((x - b)/a)$ and $u(b) = W_f(a, b)$ (with x and a constant). In order to apply the Plancherel theorem, both of these functions must belong to $L^2(R)$. If f and ψ have finite support, then the b-support of $W_f(a, b)$ will also be finite (as you can check) and so $W_f(a, b)$ and $\psi\left(\frac{x-b}{a}\right)$ are L^2 functions in b.

- **Step 2.** The next step is to evaluate the two Fourier-b-transforms occurring in the integrand of Step 1. We show

$$\overline{\mathcal{F}_b\left\{\psi\left(\frac{x-b}{a}\right)\right\}(y)} = ae^{iyx}\widehat{\psi}(ay)$$

$$\mathcal{F}_b\left\{W_f(a,b)\right\}(y) = a\sqrt{\frac{2\pi}{|a|}}\,\overline{\widehat{\psi}(ay)}\,\widehat{f}(y)$$

We establish Step 2 later. Assuming this step for the moment, we proceed to the conclusion. Combining Steps 1 and 2, we obtain

$$F(x) = \frac{1}{C_\psi}\int_{-\infty}^{\infty}\frac{\sqrt{2\pi}\,da}{|a|}\int_{-\infty}^{\infty}|\widehat{\psi}(ay)|^2\widehat{f}(y)e^{iyx}\,dy \tag{7.11}$$

$$= \frac{1}{C_\psi}\int_{-\infty}^{\infty}\sqrt{2\pi}\widehat{f}(y)e^{iyx}\,dy\int_{-\infty}^{\infty}\frac{|\widehat{\psi}(ay)|^2}{|a|}\,da \tag{7.12}$$

where the last equality follows by interchanging the order of the y- and a-integrals. To calculate the a-integral on the right, we make a change of variables $u = ay$ (and so $du = y\,da$) provided that $y \neq 0$ to obtain

$$\int_{-\infty}^{\infty}\frac{|\widehat{\psi}(ay)|^2}{|a|}\,da = \int_{-\infty}^{\infty}\frac{|\widehat{\psi}(u)|^2}{|u|}\,du \tag{7.13}$$

$$= \frac{C_\psi}{2\pi}. \tag{7.14}$$

If $y = 0$, then by assumption $\widehat{\psi}(0) = 0$ and so the rightmost integral in (7.12) is zero. Therefore, the rightmost integral in (7.12) is equal to $C_\psi/2\pi$ for all values of y except $y = 0$. Since the y-integral in (7.12) is not affected by the value of the integrand at the single point $y = 0$, we can substitute (7.14) into (7.12) to obtain

$$F(x) = \frac{1}{C_\psi}\int_{-\infty}^{\infty}\sqrt{2\pi}\widehat{f}(y)e^{iyx}\frac{C_\psi}{2\pi}\,dy$$

$$= \frac{1}{\sqrt{2\pi}}\int_{-\infty}^{\infty}\widehat{f}(y)e^{iyx}\,dy$$

$$= f(x)$$

where the last equality follows from the Fourier inversion theorem. This finishes the proof, modulo the derivation of Step 2, which we present next.

Proof of Step 2 The first equality in Step 2 follows from the sixth property of Theorem 2.6. In more detail, we use the definition of the Fourier transform to obtain

$$\mathcal{F}_b\left(\overline{\psi\left(\frac{x-b}{a}\right)}\right)(y) = \frac{1}{\sqrt{2\pi}}\int_{-\infty}^{\infty}\overline{\psi\left(\frac{x-b}{a}\right)}e^{-iby}\,db$$

After the change of variables $v = (x-b)/a$, this becomes

$$\mathcal{F}_b\left(\overline{\psi\left(\frac{x-b}{a}\right)}\right)(y) = \frac{a}{\sqrt{2\pi}}\int_{-\infty}^{\infty}\overline{\psi(v)}e^{-iy(x-va)}\,dv \qquad (7.15)$$

$$= ae^{-iyx}\overline{\frac{1}{\sqrt{2\pi}}\int_{-\infty}^{\infty}\psi(v)e^{-iayv}\,dv} \qquad (7.16)$$

$$= ae^{-iyx}\overline{\widehat{\psi}(ay)}. \qquad (7.17)$$

Taking conjugates of both sides of this equation establishes the first equality in Step 2.

To establish the second equality, we use the definition of the wavelet transform

$$\mathcal{F}_b\left\{W_f(a,b)\right\}(y) = \mathcal{F}_b\left\{\frac{1}{\sqrt{|a|}}\int_{-\infty}^{\infty}\overline{\psi\left(\frac{x-b}{a}\right)}f(x)\,dx\right\}(y).$$

The next step is to bring the Fourier transform operator $\mathcal{F}_b$ under the integral sign. This can be accomplished by viewing the integral as a sum and using the linearity of the Fourier transform operator. More precisely, we approximate the integral on the right by the following Riemann sum:

$$\mathcal{F}_b\left\{\frac{1}{\sqrt{|a|}}\int_{-\infty}^{\infty}\overline{\psi\left(\frac{x-b}{a}\right)}f(x)\,dx\right\}(y)$$

$$\approx \mathcal{F}_b\left\{\frac{1}{\sqrt{|a|}}\sum_j\overline{\psi\left(\frac{x_j-b}{a}\right)}f(x_j)\Delta x\right\}(y)$$

This approximation becomes more accurate as the partition gets finer (i.e., as $\Delta x \mapsto 0$). The Fourier transform operator, $\mathcal{F}_b$, is linear and can be brought inside the sum to obtain

$$\mathcal{F}_b\left\{\frac{1}{\sqrt{|a|}}\int_{-\infty}^{\infty}\overline{\psi\left(\frac{x-b}{a}\right)}f(x)\,dx\right\}(y)$$

$$\approx \frac{1}{\sqrt{|a|}}\sum_j\mathcal{F}_b\left\{\overline{\psi\left(\frac{x_j-b}{a}\right)}\right\}(y)\,f(x_j)\Delta x.$$

Now we let the partition get finer and obtain (in the limit as $\Delta x \mapsto 0$)

$$\mathcal{F}_b\left\{W_f(a,b)\right\}(y) = \frac{1}{\sqrt{|a|}}\int_{-\infty}^{\infty}\mathcal{F}_b\left\{\overline{\psi\left(\frac{x-b}{a}\right)}\right\}(y)f(x)\,dx.$$

Inserting (7.17) into this equation, we obtain

$$\mathcal{F}_b\left\{W_f(a,b)\right\}(y) = \frac{a}{\sqrt{|a|}} \int_{-\infty}^{\infty} e^{-iyx}\,\overline{\widehat{\psi}(ay)} f(x)\,dx$$

$$= \frac{a}{\sqrt{|a|}}\overline{\widehat{\psi}(ay)} \int_{-\infty}^{\infty} f(x)e^{-iyx}\,dx$$

$$= \frac{a\sqrt{2\pi}}{\sqrt{|a|}}\,\overline{\widehat{\psi}(ay)}\,\widehat{f}(y)$$

as claimed in the second equation of Step 2. This completes the derivation in Step 2 and the proof of the theorem.   ◆

Appendix A

Technical Matters

A.1 Proof of the Fourier Inversion Formula

In this section we give a rigorous proof of Theorem 2.1 (the Fourier inversion theorem), which states that for an integrable function f,

$$f = \mathcal{F}^{-1} \circ \mathcal{F}(f).$$

Inserting the definitions of $\mathcal{F}$ and $\mathcal{F}^{-1}$, we must show

$$f(x) = \frac{1}{2\pi} \int_{-\infty}^{\infty} \int_{-\infty}^{\infty} f(t) e^{-i(t-x)\lambda} \, dt \, d\lambda.$$

We restrict our attention to functions, f, which are nonzero only on some finite interval, to avoid the technicalities of dealing with convergent integrals over infinite intervals (for details, see Tolstov [22]). If f is nonzero only on a finite interval, then the t-integral occurs only on this finite interval (instead of $-\infty < t < \infty$ as it appears). The λ-integral still involves an infinite interval and so this must be handled by integrating over a finite interval of the form $-l \le \lambda \le l$ and then letting $l \to \infty$. So we must show

$$f(x) = \frac{1}{2\pi} \lim_{l \to \infty} \int_{-l}^{l} \int_{-\infty}^{\infty} f(t) e^{-i(t-x)\lambda} \, dt \, d\lambda.$$

Using the definition of the complex exponential ($e^{iu} = \cos u + i \sin u$), the preceding limit is equivalent to showing

$$f(x) = \frac{1}{2\pi} \lim_{l \to \infty} \int_{-l}^{l} \int_{-\infty}^{\infty} f(t) \left(\cos((t-x)\lambda) - i \sin((t-x)\lambda) \right) \, dt \, d\lambda.$$

Since sin is an odd function, the λ-integral involving $\sin((t - x)\lambda)$ is zero. Together with the fact that cos is an even function, the preceding limit is equivalent to

$$f(x) = \frac{1}{\pi} \lim_{l \to \infty} \int_0^l \int_{-\infty}^{\infty} f(t) \cos((t - x)\lambda)\, dt\, d\lambda.$$

Now $\int_0^l \cos((t - x)\lambda)\, d\lambda = \frac{\sin((t-x)l)}{t-x}$. By replacing t by $x + u$, the preceding limit is equivalent to

$$f(x) = \frac{1}{\pi} \lim_{l \to \infty} \int_{-\infty}^{\infty} f(x + u) \frac{\sin(lu)}{u}\, du. \tag{A.1}$$

To establish this limit, we must show that for any given $\epsilon > 0$, the difference between $f(x)$ and the integral on the right is less than ϵ provided that l is sufficiently large. For this given ϵ, we choose a $\delta > 0$ so that

$$\frac{1}{\pi} \int_{-\delta}^{\delta} |f(x + u)|\, du < \epsilon \tag{A.2}$$

(geometrically, the integral on the left is the area under the graph of $|f|$ over an interval of width 2δ and by choosing δ small enough, we can arrange that the area of this sliver is less than ϵ). We use this inequality at the end of our proof.

Now we need to use the Riemann-Lebesgue lemma (Theorem 1.21), which states

$$\lim_{l \to \infty} \int_a^b g(u) \sin(lu)\, du = 0$$

where g is any piecewise continuous function. Here, a or b could be infinity if g is nonzero only on a finite interval. By letting $g(u) = f(x + u)/u$, this lemma implies

$$\frac{1}{\pi} \int_{-\infty}^{-\delta} f(x + u) \frac{\sin(lu)}{u}\, du \quad \text{and} \quad \frac{1}{\pi} \int_{\delta}^{\infty} f(x + u) \frac{\sin(lu)}{u}\, du \;\to\; 0$$

as $l \to \infty$ [the function $g(u) = f(x + u)/u$ is continuous on both intervals of integration]. Thus the limit in (A.1) is equivalent to showing

$$f(x) = \frac{1}{\pi} \lim_{l \to \infty} \int_{-\delta}^{\delta} f(x + u) \frac{\sin(lu)}{u}\, du. \tag{A.3}$$

On the other hand, we know from Step 5 in the proof of the convergence theorem for Fourier series [see Equation (1.27)] that

$$\frac{1}{\pi} \lim_{n \to \infty} \int_{-\pi}^{\pi} f(x + u) \frac{\sin((n + 1/2)u)}{2\sin(u/2)}\, du = f(x) \quad \text{(here n is an integer).} \tag{A.4}$$

So our proof of (A.3) will proceed in two steps.

Step 1.   We show

$$\frac{1}{\pi} \int_{-\pi}^{\pi} f(x+u) \frac{\sin((n+1/2)u)}{2\sin(u/2)} \, du - \frac{1}{\pi} \int_{-\delta}^{\delta} f(x+u) \frac{\sin((n+1/2)u)}{u} \, du \;\to\; 0$$

as $n \to \infty$. After this step has been established, it [together with (A.4)] shows that

$$\frac{1}{\pi} \int_{-\delta}^{\delta} f(x+u) \frac{\sin((n+1/2)u)}{u} \, du \;\to\; f(x) \quad \text{as } n \to \infty, \qquad (A.5)$$

which is the same as the limit in (A.3) for l of the form $l = n+1/2$. To establish the more general limit in (A.3), we need the following.

Step 2.   Any $l > 0$ can be written as $l = n + h$, where n is an integer and $0 \le h < 1$. We show

$$\frac{1}{\pi} \int_{-\delta}^{\delta} f(x+u) \left(\frac{\sin((n+1/2)u)}{u} - \frac{\sin(lu)}{u} \right) du \; < \epsilon/2.$$

Once these steps have been established, the proof of (A.3) (and therefore the proof of the Fourier inversion theorem) is completed as follows. Using (A.5), we can choose N large enough so that if $n > N$, then

$$\left| f(x) - \frac{1}{\pi} \int_{-\delta}^{\delta} f(x+u) \frac{\sin((n+1/2)u)}{u} \, du \right| < \epsilon/2.$$

This inequality, together with the one in Step 2, yields

$$\left| f(x) - \frac{1}{\pi} \int_{-\delta}^{\delta} f(x+u) \frac{\sin(lu)}{u} \, du \right|$$

$$\le \left| f(x) - \frac{1}{\pi} \int_{-\delta}^{\delta} f(x+u) \frac{\sin((n+1/2)u)}{u} \, du \right|$$

$$+ \left| \frac{1}{\pi} \int_{-\delta}^{\delta} f(x+u) \left(\frac{\sin((n+1/2)u)}{u} - \frac{\sin(lu)}{u} \right) du \right|$$

$$< \epsilon/2 + \epsilon/2 \quad \text{if } n > N$$

and our proof is complete.

Proof of Step 1   The statement of Step 1 is equivalent to

$$\frac{1}{\pi} \int_{-\delta}^{\delta} f(x+u) \sin((n+1/2)u) \left(\frac{1}{2\sin(u/2)} - \frac{1}{u} \right) du \to 0 \quad \text{as } n \to \infty. \quad (A.6)$$

Because

$$\int_{-\pi}^{-\delta} \frac{f(x+u)}{2\sin(u/2)} \sin((n+1/2)u)\, du \quad \text{and} \quad \int_{\delta}^{\pi} \frac{f(x+u)}{2\sin(u/2)} \sin((n+1/2)u)\, du \to 0$$

as $n \to \infty$ by the Riemann-Lebesgue lemma, ($\frac{f(x+u)}{2\sin(u/2)}$ is continuous over the intervals $-\pi \le u \le -\delta$ and $\delta \le u \le \pi$). In addition, the quantity

$$\frac{1}{2\sin(u/2)} - \frac{1}{u}$$

is continuous on the interval $-\delta \le u \le \delta$ because the only possible discontinuity occurs at $u = 0$ and the limit of this expression as $u \to 0$ is zero (using L'Hôspital's rule or a Taylor expansion). Therefore, the Riemann-Lebesgue lemma implies that the limit in (A.6) holds and the derivation of Step 1 is complete.

Proof of Step 2 For any $l > 0$, we write $l = n + h$, where n is an integer and $0 \le h < 1$. Using the mean value theorem [which states that $f(x) - f(y) = f'(t)(x - y)$ for some t between x and y], we have

$$|\sin((n+1/2)u) - \sin(lu)| = |\sin((n+1/2)u) - \sin((n+h)u)|$$
$$= |\cos(t)|\,|u/2 - hu|$$
$$\le |u|/2 \quad \text{since } 0 \le h < 1.$$

Therefore,

$$\frac{1}{\pi} \int_{-\delta}^{\delta} |f(x+u)| \left(\frac{\sin((n+1/2)u)}{u} - \frac{\sin(lu)}{u} \right) |\, du$$

$$\le \frac{1}{\pi} \int_{-\delta}^{\delta} |f(x+u)| \frac{|u|}{2|u|}$$

$$\le \frac{\epsilon}{2} \quad \text{by equation (A.2).}$$

This completes the proof of Step 2 and the proof of the theorem.   ◆

A.2   Rigorous Proof of Theorem 5.17

We restate Theorem 5.17 as follows:

Suppose ϕ is a continuous function with compact support satisfying the orthonormality condition: that is, $\int \phi(x-k)\phi(x-l)\, dx = \delta_{kl}$. Let $V_j = \{f = \sum_k a_k \phi(2^j x - k);\ a_k \in R\}$. Then the following hold.

1. The spaces V_j satisfy the separation condition (i.e., $\cap V_j = \{0\}$).

2. If the following additional conditions are satisfied by ϕ

- *normalization:* $\int \phi(x)\, dx = 1$
- *scaling:* $\phi(x) = \sum_k p_k \phi(2x - k)$ for some finite number of constants p_k

then the associated V_j satisfy the density condition; that is, $\cup V_j = L^2(R)$ or, in other words, any element in $L^2(R)$ can be approximated by elements in V_j for large enough j.

In particular, if the function ϕ is continuous with compact support and satisfies the normalization, scaling, and orthonormality conditions listed previously, then the collection of spaces $\{V_j, j \in Z\}$ forms a multiresolution analysis.

Proof of Part 1 Note that the first part does not require the normalization or the the scaling conditions. The first part of the theorem is true under the following more general hypothesis: that there is a uniform constant C such that

$$\max_{x \in R} |f(x)| \le C\|f\|_{L^2} \quad \text{for all } f \in V_0. \tag{A.7}$$

Following Strichartz [21], we clarify why this is a more general hypothesis than requiring ϕ to have compact support. If $f \in V_0$, then $f(x) = \sum_k a_k \phi(x-k)$, where the a_k can be determined by taking the L^2 inner product of $f(x)$ with $\phi(x - k)$ [since the $\phi(x - k)$ are orthonormal]. We obtain

$$f(x) = \left(\sum_k \int f(y)\overline{\phi(y - k)}\, dy \right) \phi(x - k) = \int k(x, y) f(y)\, dy$$

where $k(x, y) = \sum_k \phi(x - k)\overline{\phi(y - k)}$. Using the Schwarz inequality for L^2 (often called Hölder's inequality), we obtain

$$|f(x)| \le \left(\int k(x, y)^2\, dy \right)^{1/2} \|f\|_{L^2} = \left(\sum_k |\phi(x - k)|^2 \right)^{1/2} \|f\|_{L^2}$$

where the last equality again uses the orthonormality of the $\phi(y - k)$. Since ϕ is assumed to vanish outside a finite interval, the sum on the right is over a fixed, finite number of indices and is therefore bounded above by some finite constant C. Thus estimate (A.7) is satisfied by any compactly supported ϕ whose translates are orthonormal.

Now we establish the first part of the theorem assuming the inequality in (A.7). Suppose f belongs to V_{-j}. Then $f(2^j x)$ belongs to V_0 and, using (A.7), we have

$$|f(2^j x)| \le C \left(\int |f(2^j y)|^2\, dy \right)^{1/2} = C 2^{-j/2} \left(\int |f(t)|^2\, dt \right)^{1/2}$$

where the last equality uses the change of variables $t = 2^j y$. Since the preceding inequality holds for all x, we conclude

$$\max_{x \in R} |f(x)| \leq C 2^{-j/2} \|f\|_{L^2}.$$

If f belongs to all V_{-j}, then this inequality holds for all j. Letting $j \to \infty$, we conclude that f must be zero everywhere.

Proof of Part 2   For the proof of the second part, we consider the orthogonal projection map $P_j : L^2 \to V_j$. Since $\{\phi_{jk}(x) = 2^{j/2}\phi(2^j x - k),\ k \in Z\}$ is an orthonormal basis for V_j,

$$P_j f(x) \;=\; \sum_k \langle f, \phi_{jk} \rangle \phi_{jk}(x) \tag{A.8}$$

$$=\; 2^j \sum_k \left(\int f(y)\overline{\phi(2^j y - k)}\, dy \right) \phi(2^j x - k) \tag{A.9}$$

for any $f \in L^2$. We must show that for each $f \in L^2$, $P_j f \to f$ in L^2 as $j \to \infty$. Equivalently, we must show that

$$\|P_j f\| \to \|f\| \tag{A.10}$$

(all norms are L^2 norms). This equivalence is seen as follows. Since $f = (f - P_j f) + P_j f$ and since $P_j f$ is orthogonal to $f - P_j f$ (by the definition of an orthogonal projection), we have

$$\|f\|^2 = \|f - P_j f\|^2 + \|P_j f\|^2.$$

Therefore, $\|P_j f\|^2 \to \|f\|^2$ if and only if $\|f - P_j f\|^2 \to 0$.

We establish (A.10) in three steps.

Step 1. $\|P_j f\| \to \|f\|$ for any characteristic function of the form

$$\chi(x) = \begin{cases} 1 & \text{if } a \leq x \leq b \\ 0 & \text{otherwise} \end{cases}$$

for some numbers $a < b$.

In view of (A.9),

$$P_j \chi(x) = 2^j \sum_k \left(\int_a^b \overline{\phi(2^j y - k)}\, dy \right) \phi(2^j x - k).$$

Therefore,

$$\|P_j \chi\|^2$$

$$= 2^{2j} \sum_{k,k'} \left(\int_a^b \overline{\phi(2^j y - k)}\, dy \right) \left(\int_a^b \phi(2^j y - k')\, dy \right) \int \phi(2^j x - k)\overline{\phi(2^j x - k')}\, dx$$

$$= 2^j \sum_k \left| \int_a^b \phi(2^j y - k)\, dy \right|^2$$

where the last equality uses the orthonormality of the $2^{j/2}\phi(2^j x - k)$ in $L^2(R)$. Using the change of variables, $t = 2^j y$,

$$\|P_j \chi\|^2 = 2^{-j} \sum_k \left| \int_{2^j a}^{2^j b} \phi(t - k)\, dt \right|^2.$$

When j is large, the interval of integration $2^j a \leq t \leq 2^j b$ is very large and the support of ϕ (i.e., essentially the set where ϕ is nonzero) is small by comparison. Therefore, except for a few indices k near the boundary of this interval, the indices fall into two categories: (1) The support of $\phi(t - k)$ does not overlap the interval of integration $2^j a \leq t \leq 2^j b$, in which case the integral on the right is zero; (2) the support of $\phi(t - k)$ is totally contained inside the interval of integration $2^j a \leq t \leq 2^j b$, in which case the integral on the right is 1 due to the normalization condition $\int \phi(y)\, dy = 1$. Therefore,

$$\|P_j \chi\|^2 \approx 2^{-j} \left(\text{the number of integers between } 2^j a \text{ and } 2^j b\right)$$
$$\approx (b - a)$$
$$= \int_a^b 1\, dy$$
$$= \|\chi\|^2.$$

The error in these approximations is due to the few indices k when the support of $\phi(x - k)$ overlaps the boundary of the interval $2^j a \leq x \leq 2^j b$. As j gets large, the number of these indices gets smaller in comparison to the number of integers between $2^j a$ and $2^j b$. Therefore, the approximations get more accurate as $j \to \infty$ and so Step 1 is complete.

Step 2. $\|P_j s\| \to \|s\|$ as $j \to \infty$, where s is a step function; that is, a finite sum of the form

$$s = \sum_k \alpha_k \chi_k.$$

Here, each χ_k is a characteristic function of the type discussed in Step 1 (see Figure 1 for a graph of a typical step function).
We have

$$\|P_j(s) - s\| = \left\| \sum_k \alpha_k (P_j(\chi_k) - \chi_k) \right\| \leq \sum_k |\alpha_k|\, \|P_j(\chi_k) - \chi_k\|.$$

Step 1 established $\|P_j(\chi_k) - \chi_k\| \to 0$ as $j \to \infty$ for each k. Since there are only a finite number of indices, clearly $\|P_j s - s\| \to 0$ as well.

Step 3. $P_j f \to f$ for a general $f \in L^2$. This follows from Step 2 because an arbitrary function $f \in L^2$ can be approximated by step functions, as illustrated in Figure 1.

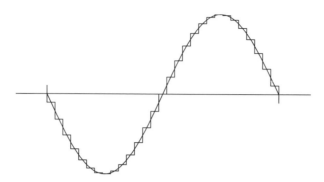

Figure 1 Approximating by step functions

We proceed as follows. For a given ϵ, choose a step function s with $\|f - s\| < \epsilon/3$. Since an orthogonal projection operator does not increase norm, $\|P_j(f-s)\|$ is also less than $\epsilon/3$. From Step 2, $\|P_j s - s\| < \epsilon/3$ for j large enough. Thus for large j,

$$\|f - P_j(f)\| = \|(f - s) + (s - P_j s) + (P_j s - P_j f)\|$$
$$\leq \|f - s\| + \|s - P_j s\| + \|P_j(s - f)\|$$
$$< \epsilon/3 + \epsilon/3 + \epsilon/3.$$

Step 3 and the proof of the second part of the theorem are now complete.

A.2.1   Proof of Theorem 5.10

In Definition 5.1 (Chapter 5), we defined V_1 to be the linear space spanned by $\phi(2x - j)$, $j \in Z$, where ϕ is a scaling function satisfying

$$\phi(x) = \sum_l p_l \phi(2x - l) \tag{A.11}$$

for some finite number of constants p_k. The associated wavelet function is defined as

$$\psi(x) = \sum_l (-1)^l \overline{p_{1-l}} \phi(2x - l). \tag{A.12}$$

In the proof of Theorem 5.10 given earlier, we have shown that the translates of ψ are orthogonal to the translates of ϕ. The goal of this section is to show that V_1 is spanned by the translates of ϕ and the translates of ψ. This will imply that any element in W_0 (the orthogonal complement of V_0 in V_1) is a linear combination of the translates of ψ. This is the missing ingredient in our earlier proof of Theorem 5.10.

We must show that for each j

$$\phi(2x - j) = \sum_k a_k \phi(x - k) + b_k \psi(x - k)$$

for some choice of constants a_k and b_k. Since $\{\phi(x-k),\ k \in Z\}$ is orthonormal, the constants a_k, if they exist, must be given by

$$
\begin{aligned}
a_k &= \int \phi(2y - j)\overline{\phi(y - k)}\, dy \\
&= \int \sum_l \phi(2y - j)\overline{p_l}\,\overline{\phi(2y - 2k - l)} \quad \text{by (A.11)} \\
&= (1/2)\overline{p_{j-2k}}
\end{aligned}
$$

where the last equality again uses the fact that $\{2^{(1/2)}\phi(2y - j);\ j \in Z\}$ is orthonormal in L^2. Similarly, the constants b_k, if they exist, must be given by

$$
b_k = 2^{-1}(-1)^j p_{1-j+2k}.
$$

So we must show

$$
\phi(2x - j) = 2^{-1}\sum_k \overline{p_{j-2k}}\,\phi(x - k) + (-1)^j p_{1-j+2k}\psi(x - k).
$$

Using (A.11) and (A.12), this equation is equivalent to

$$
\phi(2x - j) = (1/2)\sum_{k,l} \left((-1)^{j+l}p_{1-j+2k}\overline{p_{1-l}} + \overline{p_{j-2k}}p_l\right)\phi(2x - 2k - l).
$$

In order for this equation to hold, the coefficient of $\phi(2x - 2k - l)$ must be zero unless $2k + l = j$. Therefore, we must show that if $l = j - 2k$, then

$$
\sum_k p_{1-j+2k}\overline{p_{1-j+2k}} + \overline{p_{j-2k}}p_{j-2k} = 2 \tag{A.13}
$$

and if $l = j - 2k + t$ for $t \neq 0$, then

$$
\sum_k (-1)^t p_{1-j+2k}\overline{p_{1-j+2k-t}} + \sum_k \overline{p_{j-2k}}p_{j-2k+t} = 0 \tag{A.14}
$$

for all $t \neq 0$.

By letting $\gamma = 1 - j + 2k$ and then $\gamma = j - 2k$, the left side of (A.13) can be condensed into

$$
\sum_\gamma \overline{p_\gamma}p_\gamma,
$$

which equals 2 by Theorem 5.9, as desired.

For (A.14), we need to handle the cases when t is odd and even separately. If t is odd, say $t = 2s + 1$, then the first sum in (A.14) can be transformed by a change of index ($k' = s + j - k$) into

$$
\sum_{k'} -p_{j-2k'+t}\overline{p_{j-2k'}},
$$

which cancels with the second sum. If t is even, say $t = 2s$, then the change of index $k = -k' + j + s$ brings the second sum to the form

$$\sum_{k'} p_{-j+2k'} \overline{p_{-j+2k'-t}}.$$

By letting $\gamma = -j + 2k'$ for this sum and then $\gamma = 1 - j + 2k$ for the first sum in (A.14), the two sums in (A.14) can be combined as

$$\sum_{\gamma} p_{\gamma} \overline{p_{\gamma-2s}},$$

which equals zero by Theorem 5.9. The proof of the theorem is now complete.

A.2.2   Proof of the Convergence Part of Theorem 5.23

In this section, we prove the convergence part of the algorithm described in Theorem 5.23 used to construct the scaling function ϕ from the scaling equation

$$\phi(x) = \sum_{k} p_k \phi(2x - k).$$

Recall that this algorithm starts by setting ϕ_0 equal to the Haar scaling function. Then we set

$$\phi_n(x) = \sum_{k} p_k \phi_{n-1}(2x - k)$$

for $n = 1, 2, \ldots$. As explained in Section 5.3.3 on Fourier transform criteria, the Fourier transform of this equation is

$$\widehat{\phi_n}(\xi) = P(e^{-i\xi/2})\widehat{\phi_{n-1}}(\xi/2).$$

Iterating this equation $n - 1$ times yields

$$\widehat{\phi_n}(\xi) = \prod_{j=1}^{n-1} P(e^{-i\xi/2^j})\widehat{\phi_0}(\xi/2^n).$$

Our goal in this section is to show that the sequence ϕ_n defined inductively by this procedure converges to a function ϕ in L^2. We do this in two steps. We first show that $\widehat{\phi_n}$ converges uniformly on the compact subsets of R. Then we show that ϕ_n converges in the L^2 norm. As shown in Section 5.3.4, each ϕ_n satisfies the orthonormality condition. If ϕ_n and all its translates are orthonormal, then the same must be true of its L^2 limit.

For the first step, we need the following lemma.

LEMMA A.1   *If e_j is a sequence of functions such that $\sum_j |e_j|$ converges uniformly on a set K, then the infinite product $\prod_j (1 + e_j)$ also converges uniformly on the set K.*

A sketch of the proof of the lemma is as follows. Write the product as

$$\prod_j (1 + e_j) = e^{\ln\left(\Pi_j (1+e_j)\right)} = e^{\Sigma_j \ln(1+e_j)}.$$

If the series $\sum_j |e_j|$ converges, then e_j must converge to zero. When $|x|$ is small, $|\ln(1 + x)| \approx |x|$ (since by L'Hôspital's rule $|\ln(1 + x)|/|x| \to 1$ as $|x| \to 0$). Therefore, the uniform convergence of $\sum_j |e_j|$ on K is equivalent to the uniform convergence of $\sum_j |\ln(1 + e_j)|$ which in turn implies the uniform convergence of $\prod_j (1 + e_j)$ by the preceding equalities.

We want to apply the preceding lemma to the case $e_j = p(\xi/2^j) - 1$ where $p(\xi) = P(e^{-i\xi})$ and where K is any compact subset of R. Since p is differentiable (it is a polynomial), and since $p(0) = 1$, we have $|p(t) - 1| \leq C|t|$. Thus $|e_j| \leq C|\xi/2^j|$ and so $\sum_j |e_j|$ converges uniformly on each compact subset of R. By the lemma,

$$\widehat{\phi_n}(\xi) = \prod_{j=1}^{n-1} p(\xi/2^j)\widehat{\phi_0}(\xi/2^n) \tag{A.15}$$

also converges as $n \to \infty$ uniformly on each compact set in R to a function we call

$$g(\xi) = \prod_{j=1}^{\infty} p(\xi/2^j)\widehat{\phi_0}(0).$$

Now $\|\widehat{\phi_0}\|_{L^2} = 1$ (by construction) and $|p(\xi)| \leq 1$ (by hypothesis). Therefore $\|\widehat{\phi_n}\|_{L^2} \leq 1$ for all n. By Fatou's lemma, $\int \lim_{n\to\infty} |\widehat{\phi_n}|^2 \leq \liminf_{n\to\infty} \int |\widehat{\phi_n}|^2 \leq 1$ and so g belongs to L^2. By Theorem 2.1, g equals $\widehat{\phi}$ for some ϕ in L^2.

We now show the sequence $\phi_n \to \phi$ in L^2, or equivalently, that $\widehat{\phi_n} \to \widehat{\phi}$ in L^2. To this end, we use the dominated convergence theorem, which states that if a sequence of functions f_n converges pointwise to a limit function, f, and if there is an integrable function F that dominates the sequence (i.e., $|f_n| \leq |F|$ for all n) then $\int f_n \to \int f$. In our context, the $\widehat{\phi_n}$ converges pointwise to $\widehat{\phi}$ (in fact uniformly on each compact set). So we need only dominate all the $\widehat{\phi_n}$ by a single function belonging to L^2.

Now, by assumption, $P(e^{-i\xi}) \neq 0$ on the interval $-\pi/2 \leq \xi \leq \pi/2$. Therefore, $P(e^{-i\xi/2^j}) \neq 0$ on the interval $-2^{j-1}\pi \leq \xi \leq 2^{j-1}\pi$ and so

$$\widehat{\phi}(\xi) = \prod_{j=1}^{\infty} p(\xi/2^j)\widehat{\phi_0}(0) \neq 0 \tag{A.16}$$

on $-\pi/2 \leq \xi \leq \pi/2$. Since this function is continuous, it is bounded away from zero (i.e., $\widehat{\phi} \geq c$ on this compact set). This inequality implies

$$|\widehat{\phi}(\xi/2^{n-1}))| = \prod_{j=n}^{\infty} |p(\xi/2^j)\widehat{\phi_0}(0)| \geq c \text{ for } |\xi| \leq 2^{n-2}\pi. \tag{A.17}$$

Using (A.15), (A.16), and (A.17),

$$|\widehat{\phi_n}(\xi)| = \frac{\prod_{j=1}^{\infty} |P(e^{-i\xi/2^j})\widehat{\phi_0}(\xi/2^n)|}{\prod_{k=n}^{\infty} |P(e^{-i\xi/2^k})|}$$

$$= \frac{|\widehat{\phi}(\xi)|}{|\widehat{\phi}(\xi/2^n)|} |\widehat{\phi_0}(\xi/2^n)|$$

$$\leq \frac{1}{c} |\widehat{\phi}(\xi)| \widehat{\phi_0}(\xi/2^n).$$

Now

$$\widehat{\phi_0}(\xi) = \frac{e^{-i\xi-1}}{-\sqrt{2\pi}\, i\xi},$$

as computed in (5.27). Note that $|\widehat{\phi_0}(\xi)| \to 0$ as $|\xi| \to \infty$ and $\widehat{\phi_0}(\xi)$ is bounded as $|\xi| \to 0$ (by L'Hôpital's Rule). Therefore $|\widehat{\phi_0}(\xi)|$ is a bounded function, and we have

$$|\widehat{\phi_n}(\xi)| \leq C|\widehat{\phi}(\xi)|$$

for some uniform constant C. Since $\widehat{\phi}$ has already been shown to be an element of L^2, the expression on the right can be used as our dominating function for use in the dominated convergence theorem for the sequence $\widehat{\phi_n}$. Thus, $\widehat{\phi_n} \to \widehat{\phi}$ and so $\phi_n \to \phi$ in L^2 as $n \to \infty$. The proof of the L^2 convergence of our algorithm is complete.

Appendix B

MATLAB Routines

B.1   General Compression Routine

The following MATLAB function compresses a given vector-zeroing out the terms
falling below a user specified (percentage) threshold. This routine is used by all
FFT and wavelet compression schemes that follow.

```
function wc=compress(w,r)
% Input is the array w and r, which is a number
% strictly between 0 and 1.
% Output is the array wc where smallest 100r% of the
% terms in w are set to zero (e.g r=.75 means the smallest 75% of
% the terms are set to zero
if (r<0) | (r>1)
  error('r should be between 0 and 1')
end;
N=length(w); Nr=floor(N*r);
ww=sort(abs(w));
tol=abs(ww(Nr+1));
wc=(abs(w)>=tol).*w;
```

B.2   Use of MATLAB's FFT Routine for Filtering and Compression

Filtering with FFT.   The following MATLAB commands are needed for
Example 3.6, on filtering the high-frequency components from a given signal.

The key MATLAB commands are fft and ifft, which compute the fast Fourier transform and its inverse, respectively.

```
>> t=linspace(0,2*pi,2^8); % discretizes [0, 2pi] into 256 nodes
>> y=exp(-(cos(t).^2)).*(sin(2*t)+2*cos(4*t)
     +0.4*sin(t).*sin(50*t));
>> plot(t,y) % generates the graph of the original signal
>> fy=fft(y); % computes fft of y
>> filterfy=[fy(1:6) zeros(1,2^8-12) fy(2^8-5:2^8)]; % sets fft
>>% coefficients to zero for $7 \leq k \leq 256$
>> filtery=ifft(filterfy); % computes inverse fft of the filtered
>>         % fft
>> plot(t, filtery)  % generates the plot of the compressed signal
```

Compression with the FFT. The following routine, called fftcomp.m, uses the FFT and the previous compress.m routine to compress a given signal. The compression rate is part of the input. The routine outputs the graph of the signal, the graph of the compressed signal, and the relative l^2 error.

```
function error=fftcomp(t,y,r)
% Input is an array y, which represents a digitized signal
% associated with the discrete time array t.
% Also input r which is a decimal
% number between 0 and 1 representing the compression rate
% e.g. 80 percent would be r=0.8.
% Outputs are the graph of y and its compression as well as
% the relative error. This routine uses compress.m
%
if (r<0) | (r>1)
  error('r should be between 0 and 1')
end;
fy=fft(y);
fyc=compress(fy,r);
yc=ifft(fyc);
plot(t,y,t,yc)
error=norm(y-yc,2)/norm(y)
```

Use of fftcomp.m for Example 3.7 on compression with fft

```
>> t=linspace(0,2*pi,2^8);
>> y=exp(-t.^2/10).*(sin(2*t)+2*cos(4*t)+0.4*sin(t).*sin(10*t));
>> fftcomp(t,y,0.8) % uses fftcomp with compression rate of
%                        80 percent
```

B.3   Sample Routines Using MATLAB's Wavelet Toolbox

MATLAB commands needed for Compression and Filtering with Wavelets. The following MATLAB routine decomposes a signal into a wavelet decomposition using MATLAB's wavelet package. The routine compresses this decomposition and then reconstructs the compressed signal. This routine uses two key MATLAB commands, `wavedec` and `waverec`, for the decomposition and reconstruction. Different wavelets can be used with this routine. We use Daubechies 4-coefficient wavelets (indicated by the parameter `db2`). Higher-order wavelets can be used (denoted `dbn` where n is an integer from 1 to 50—$n = 1$ denotes Haar wavelets). Inputs for this routine are the signal, y, the associated time nodes, t, the number of levels, n, in the discretization, and the compression rate, r. The routine outputs the graphs of the signal and the compressed signal as well as the relative l^2 error.

```
function error=daubcomp(t,y,n,r)
% Input is an array y, which represents a digitized signal
% associated with the vector t; n=the number of levels
% (so the number of nodes is 2^n = length of t and the length
% of y).
% Also input r which is a decimal
% number between 0 and 1 representing the compression rate
% e.g. 80 percent would be r=0.8.
% Output is the graphs of y and its compression, as well as
% the relative error. This routine uses compress.m
% and the Daubechies - 4 wavelets.
%
if (r<0) | (r>1)
  error('r should be between 0 and 1')
end;
[c,l]=wavedec(y,n,'db2');  % Matlab's wave decomposition routine
cc=compress(c,r);          % compress the signal
%                            (compress.m given above)
yc=waverec(cc,l,'db2');    % Matlab's wave reconstruction
%                            routine
plot(t,y,t,yc)             % plot of the signal and compressed
%                            signal
error=norm(y-yc,2)/norm(y) % relative l^2 error
```

MATLAB commands for Figure 14

```
>> t=linspace(0,1,2^8);   % discretizes [0,1] into 256 nodes
>> y=sin(2*pi*t)+cos(4*pi*t)+sin(8*pi*t)
   +4*64*(t-1/3).*exp(-((t-1/3)*64).^2)
   +512*(t-2/3).*exp(-((t-2/3)*128).^2);
>> daubcomp(t,y,8,0.8)
```

The same routine with db2 replaced by db1 (for Haar) is used for Example 4.15.

B.4   MATLAB Code for the Algorithms in Section 5.2

The following MATLAB routine (called dec) will take a given signal, as input and return the plot of the V_j-component of the signal, where j is prescribed by the user. The wavelet decomposition down to level j is also returned. This is not intended to be a professional code (the MATLAB wavelet toolbox provides professional code). Rather, the intent here is to show how MATLAB can be used to encode the decomposition and reconstruction algorithms given in Section 5.2.

Decomposition

```
function w=dec(f,p,NJ, Jstop)
%Inputs: f = data whose length is 2^NJ, where NJ=number of scales.
%        p = scaling coefficients
%        Jstop = stopping scale; program will decompose down
%   to scale level Jstop.
%Outputs: w=wavelet coefficients down to level W-Jstop
%         the first 1:2^Jstop entries of w is the V-Jstop
%         projection
%         of f. The rest are the wavelet coefficients.
N=length(f); N1=2^NJ;
if ~(N==N1)
   error('Length of f should be 2^NJ')
end;
if (Jstop <1)|(Jstop>NJ)
   error('Jstop must be at least 1 and <= NJ')
end;
L=length(p);
pf=fliplr(p);
q=p; q(2:2:L) = -q(2:2:L);
a=f;
t=[];
for j=NJ:-1:Jstop+1
n=length(a);
a=[a(mod((-L+1:-1),n)+1) a]; % make the data periodic
b=conv(a,q);   b=b(L+1:2:L+n-1)/2;
a=conv(a,pf);   a=a(L:L+n-1)/2; % convolve
        ab=a(1:L); a=[a(L+1:n) ab];    % periodize
        a=a(2:2:n);                % then down-sample
t=[b,t];
end;
```

```
w=[a,t]; JJ=2^(Jstop);
ww=[w(JJ) w(1:JJ)];        % returns a periodic graph
tt=linspace(0,1,JJ+1);
if L==2  % for Haar, the following plot routine returns
%            a block graph
     ll=length(tt);
    ta=[tt; tt]; tt=ta(1:2*ll);
    wa=[ww; ww]; ww=wa(1:2*ll);
    ww=[ww(2*ll) ww(1:2*ll-1)];
end;
plot(tt,ww)
```

Here is the MATLAB session that generates Figure 11 on the V_4-component of a given signal.

```
>> t=linspace(0,1,2^8);  % discretize the unit interval into 2^8
%                            nodes
>> y=sin(2*pi*t)+cos(4*pi*t)+sin(8*pi*t)
   +4*64*(t-1/3).*exp(-((t-1/3)*64).^2)
   +512*(t-2/3).*exp(-((t-2/3)*128).^2); % Sample signal
>> p=[0.6830 1.1830 0.3170 -0.1830] % Coefficients for
%                             Daubechies -4.
>> w=dec(y,p,8,4);  % decomposes the signal y from level 8 down
%                        to level 4
```

Reconstruction

The following code takes the wavelet decomposition of a given signal down to level j (where j is prescribed by the user) and reconstructs the signal to top level.

```
function y=recon(w,p,NJ, Jstart)
%Inputs: w = wavelet coefficients length is 2^NJ, where
%            NJ=number of scales.
%        ordered by resolution (from lowest to highest).
%        p = scaling coefficients
%        Jstart = starting scale; program will reconstruct
%                starting with V_Jstart and ending with NJ
%
%Outputs: y=reconstructed signal at V_NJ with a corresponding
%            plot
%
N=length(w); Nj=(2^Jstart);
if ~(N==2^NJ)
  error('Length of w should be 2^NJ')
end;
if (Jstart <1)|(Jstart>NJ)
  error('Jstop must be at least 1 and <= NJ')
```

```
end;
L=length(p);
q=fliplr(p);
a=w(1:Nj);
for j=Jstart:(NJ-1)
b=w(Nj+1:2*Nj);
m=mod((0:L/2-1),Nj)+1;
Nj=2*Nj;
ua(2:2:Nj+L)=[a a(1,m)]; % periodize the data and upsample
ub(2:2:Nj+L)=[b b(1,m)]; % periodize the data and upsample
ca=conv(ua,p); ca=[ca(Nj:Nj+L-1) ca(L:Nj-1)]; % convolve with p
cb=conv(ub,q); cb=cb(L:Nj+L-1); % convolve with q
cb(1:2:Nj)=-cb(1:2:Nj); % sign change on the odd entries
a=ca+cb;
end;
y=a;
yy=[y(N) y]; % periodize the data
t=linspace(0,1,N+1);
if L==2  % in the Haar case, return a block-style graph
    ll=length(t);
    ta=[t; t]; t=ta(1:2*ll);
    ya=[yy; yy]; yy=ya(1:2*ll);
    yy=[yy(2*ll) yy(1:2*ll-1)];
end;
plot(t,yy)
```

The following MATLAB session compresses the signal f using 80% compression and reproduces Figure 14 in Chapter 5.

```
>> wc=compress(w,0.8);
>> t=linspace(0,1,2^8);  % discretize the unit interval into 2^8
%                           nodes
>> y=sin(2*pi*t)+cos(4*pi*t)+sin(8*pi*t)
  +4*64*(t-1/3).*exp(-((t-1/3)*64).^2)
  +512*(t-2/3).*exp(-((t-2/3)*128).^2); % Sample signal
>> p=[0.6830 1.18300.3170 -0.1830] % Coefficients for
%                           Daubechies -4.
>> w=dec(y,p,8,1);  %decomposes the signal y from level 8 down
%                    to level 1
>> wc=compress(w,0.8); %compresses the wavelet coefficients by
%                       80 percent
>> %(compress is the routine given at the beginning of this
%    section)
>> yc=recon(wc,p,8,1); % reconstructs from level 1 to level 8
%                         from the compressed wavelet coefficients wc
```

Bibliography

[1] Benedetto, J. J., *Harmonic Analysis and Applications*, CRC Press, Boca Raton, FL, 1997.

[2] Benedetto, J. J. and M. Frazier editors, *Wavelets: Mathematics and Applications*, CRC Press, Boca Raton, FL, 1993.

[3] Boyce, W. E. and R. C. DiPrima, *Elementary Differential Equations and Boundary Value Problems*, 3rd edition, John Wiley & Sons, Inc., New York, 1977.

[4] Burrus, S., R. Gopinath, and H. Guo, *Introduction to Wavelets and Wavelet Transforms, A Primer*, Prentice Hall, Upper Saddle River, NJ, 1998.

[5] Chui, C., *An Introduction to Wavelets, Volumes 1 and 2*, Academic Press, San Diego, 1992.

[6] Daubechies, I., *Ten Lectures on Wavelets*, SIAM, 1992.

[7] Folland, G. B., *Fourier Analysis and Its Applications*, Wadsworth & Brooks/Cole, Pacific Grove, CA, 1992.

[8] Hanselman, D. and B. Littlefield, *Mastering MATLAB 5: A Comprehensive Tutorial and Reference*, Prentice Hall, Upper Saddle River, NJ, 1998.

[9] Hernandez, E. and G. Weiss, *A First Course on Wavelets*, CRC Press, Boca Raton, FL, 1996.

[10] Mallat, S., "A theory for multi-resolution approximation: the wavelet approximation," *IEEE Trans. PAMI* **11** (1989), 674-693.

[11] Mallat, S., *A Wavelet Tour of Signal Processing*, Academic Press, San Diego, 1998.

[12] Marchuk, G. I., *Methods of Numerical Mathematics,* Springer-Verlag, Berlin, 1975.

[13] Meyer, Y., *Wavelets & Applications,* Society for Industrial and Applied Mathematics, Philadelphia, PA, 1993.

[14] Papoulis, A., *The Fourier Integral and its Applications,* McGraw-Hill, New York, 1962.

[15] Pratap, R., *Getting started with MATLAB 5,* Oxford University Press, New York, 1999.

[16] Ralston, A. and P. Rabinowitz, *A First Course in Numerical Analysis,* McGraw-Hill, New York, 1978.

[17] Royden, H., *Real Analysis, 2nd edition,* MacMillan, New York, 1968.

[18] Rudin, W. *Real and Complex Analysis,.*

[19] Stein, E. M. and G. Weiss, *Fourier Analysis on Euclidean Spaces,* Princeton University Press, Princeton, New Jersey, 1971.

[20] Strang, G. and T. Nguyen, *Wavelets and Filter Banks,* Wellesley-Cambridge Press, Wellesley, MA, 1996.

[21] Strichartz, R. S., "How To Make Wavelets," *Amer. Math. Monthly,* **100** (1993), 539-556.

[22] Tolstov, G. P., *Fourier Series,* Dover, New York, 1962.

Index

NOTES

NOTES

NOTES

NOTES

NOTES

NOTES